Advances in Anatomy
Embryology and Cell Biology

Vol. 176

Editors
F. Beck, Melbourne   B. Christ, Freiburg
W. Kriz, Heidelberg   E. Marani, Leiden
W. Kummer, Gießen   R. Putz, München
Y. Sano, Kyoto   T. H. Schiebler, Würzburg
K. Zilles, Düsseldorf

Springer-Verlag Berlin Heidelberg GmbH

Reviews and critical articles covering the entire field of normal anatomy (cytology, histology, cyto- and histochemistry, electron microscopy, macroscopy, experimental morphology and embryology and comparative anatomy) are published in Advances in Anatomy, Embryology and Cell Biology. Papers dealing with anthropology and clinical morphology that aim to encourage cooperation between anatomy and related disciplines will also be accepted. Papers are normally commissioned. Original papers and communications may be submitted and will be considered for publication provided they meet the requirements of a review article and thus fit into the scope of "Advances". English language is preferred, but in exceptional cases French or German papers will be accepted.

It is a fundamental condition that submitted manuscripts have not been and will not simultaneously be submitted or published elsewhere. With the acceptance of a manuscript for publication, the publisher acquires full and exclusive copyright for all languages and countries.

Twenty-five copies of each paper are supplied free of charge.

Manuscripts should be addressed to

Prof. Dr. F. **BECK,** Howard Florey Institute, University of Melbourne, Parkville, 3000 Melbourne, Victoria, Australia

Prof. Dr. B. **CHRIST**, Anatomisches Institut der Universität Freiburg, Abteilung Anatomie II, Albertstr. 17, D-79104 Freiburg, Germany

Prof. Dr. W. **KRIZ,** Anatomisches Institut der Universität Heidelberg, Im Neuenheimer Feld 307, D-69120 Heidelberg, Germany

Prof. Dr. W. **KUMMER,** Institut für Anatomie und Zellbiologie, Universität Gießen, Aulweg 123, D-35385 Gießen, Germany

Prof. Dr. E. **MARANI**, Leiden University, Department of Physiology, Neuroregulation Group, P.O. Box 9604, 2300 RC Leiden, The Netherlands

Prof. Dr. R. **PUTZ**, Anatomische Anstalt der Universität München, Lehrstuhl Anatomie I, Pettenkoferstr. 11, D-80336 München, Germany

Prof. Dr. Dr. h.c. Y. **SANO,** Department of Anatomy, Kyoto Prefectural University of Medicine, Kawaramachi-Hirokoji, 602 Kyoto, Japan

Prof. Dr. Dr. h.c. T. H. **SCHIEBLER**, Anatomisches Institut der Universität, Koellikerstraße 6, D-97070 Würzburg, Germany

Prof. Dr. K. **ZILLES**, Universität Düsseldorf, Medizinische Einrichtungen, C. u. O. Vogt-Institut, Postfach 101007, D-40001 Düsseldorf, Germany

D.L. Stocum

Tissue Restoration Through Regenerative Biology and Medicine

With 3 Figures

Springer

David L. Stocum, M.D.

Department of Biology and
Indiana University Center
for Regenerative Biology and Medicine
School of Science
Indiana University – Purdue University Indianapolis
402 N. Blackford St.
Indianapolis, IN 46202
USA

e-mail: dstocum@iupui.edu

ISBN 978-3-540-20603-3   ISBN 978-3-642-18928-9 (eBook)
DOI 10.1007/978-3-642-18928-9
Library of Congress Cataloging-in-Publication Data

Stocum, David L.
Tissue restoration through regenerative biology and medicine / D.L. Stocum.
    p. cm. – (Advances in anatomy, embryology, and cell biology; 176)
Includes bibliographical references and index.

1. Regeneration (Biology) 2. Transplantation of organs, tissues, etc. 3. Neural stem cells–Regeneration. I. Title. II. Series.

springeronline.com

© Springer-Verlag Berlin Heidelberg 2004
Originally published by Springer-Verlag Berlin Heidelberg New York in 2004

Preface

Regenerative biology and medicine is a rapidly developing field that has emerged from research in cell and developmental biology. Regenerative biology seeks to understand the molecular mechanisms that distinguish regeneration from fibrosis by comparative analyses of regeneration-competent vs regeneration-deficient tissues, using a wide range of experimental systems, from invertebrates such as planaria through amphibians to mammals. Regenerative medicine seeks to use this understanding to induce the regeneration of regeneration-deficient tissues. Currently, we are far from the depth of understanding of natural regeneration required for its translation into a regenerative medicine. Thus, the aim of this monograph is three-fold: (1) to summarize the current state of knowledge about the mechanisms of regeneration in some selected mammalian tissues; (2) to summarize the results of efforts that have been made to restore tissue structure and function based on our knowledge of these mechanisms; and (3) to discuss strategies to define the molecular requirements for the chemical induction of regeneration. The monograph is based on a series of lectures that comprise the graduate course "Regenerative Biology and Medicine" that I give at Indiana University-Purdue University Indianapolis and is a short version of part of a book that is in preparation.

Indianapolis, 20 November 2003   David L. Stocum

Contents

1	**Introduction**	1
1.1	The Biology of Regeneration	3
1.2	Regeneration of Ectodermal Derivatives	4
1.2.1	Uninjured Epidermis	4
1.2.2	Injured Epidermis	5
1.2.3	Hair	7
1.2.4	Nervous System	8
1.2.4.1	Axon Regeneration in Mammals....................	9
1.2.4.2	Regeneration of Amphibian and Avian Optic Nerves	13
1.2.4.3	Axon Regeneration in the Amphibian Spinal Cord .	14
1.2.4.4	Neurogenesis in the Regenerating Amphibian Tail..	16
1.2.4.5	Neurogenesis in the Adult Mammalian Brain.......	18
1.2.4.6	Injury-Induced Neurogenesis	20
1.3	Regeneration of Endodermal Derivatives	21
1.3.1	Liver...	21
1.3.1.1	Proliferation Capacity and Kinetics................	22
1.3.1.2	Gene Activity of Proliferation	23
1.3.1.3	Initiation of Liver Regeneration	24
1.3.1.4	Stopping Proliferation	26
1.3.1.5	Remodeling	26
1.3.1.6	Liver Regeneration via Stem Cells	27
1.3.2	Pancreas	27
1.4	Regeneration of Mesodermal Derivatives	28
1.4.1	Skeletal Muscle	29
1.4.1.1	Origin of Regenerated Myofibers	29
1.4.1.2	Cellular and Molecular Events of Muscle Regeneration	30
1.4.1.3	Regulation of Muscle Regeneration by Growth Factors................................	32
1.4.2	Bone..	33
1.4.2.1	Bone Regenerates via Mesenchymal Stem Cells.....	33
1.4.2.2	Regulation of Bone Regeneration by Growth Factors	34
1.4.3	Blood and Lymphoid Cells	36
1.5	Developmental Potential of Adult Stem Cells	37
1.5.1	Neural Stem Cells................................	38
1.5.2	Hepatic Oval Cells	39
1.5.3	Satellite Cells	40
1.5.4	Bone Marrow Cells	40

1.5.4.1 Unfractionated Bone Marrow 40
1.5.4.2 Hematopoietic Stem Cells........................ 41
1.5.4.3 Mesenchymal Stem Cells.......................... 42
1.5.4.4 Multipotential Adult Progenitor Cells
        of Bone Marrow.................................. 43
1.5.5   What Is the Basis of Adult Stem Cell Potency? 44

2   Regenerative Medicine 47
2.1   Cell Transplants................................ 47
2.1.1   Central Nervous System 49
2.1.1.1 Demyelinating Disorders 49
2.1.1.2 Spinal Cord Injury 50
2.1.1.3 Parkinson's Disease 50
2.1.2   Myocardial Infarction 51
2.1.3   Osteogenesis Imperfecta......................... 53
2.1.4   Diabetes 53
2.1.5   Skin Injuries................................... 54
2.1.6   Cartilage and Bone Injuries..................... 55
2.1.7   Research Issues in Cell Transplantation.......... 55
2.1.7.1 Cell Sources.................................... 55
2.1.7.2 Biomaterials for Bioartificial Tissues 56
2.2   Chemical Induction of Regeneration In Vivo 57
2.2.1   Evidence for Latent Regenerative Capacity
        in Mammals 57
2.2.1.1 New Cardiomyocytes Appear
        After Myocardial Infarction..................... 57
2.2.1.2 Stem Cells Reside in Non-regenerating Tissues 59
2.2.1.3 Induction of Dedifferentiation of Mouse Myotubes 59
2.2.2   Interventions to Promote Regeneration In Vivo.... 60
2.2.2.1 Peripheral Nerve 60
2.2.2.2 Spinal Cord Injury 61
2.2.2.3 Parkinson's Disease 65
2.2.2.4 Bone and Cartilage 65
2.2.2.5 Skin Wounds.................................... 66
2.2.2.6 Use of Pig Small Intestine and Urinary Bladder
        Submucosa as a Regeneration Template 67
2.2.3   Strategies to Define the Molecular Requirements
        for Chemical Induction of Regeneration.......... 67
2.2.3.1 Comparison of Mutant Vs Wild-Type Tissues...... 68
2.2.3.2 Comparison of Regeneration-Competent Vs
        Regeneration-Deficient Stages of the Life Cycle 69
2.2.3.3 Regeneration-Competent Vs
        Regeneration-Deficient Species 73

3   Perspectives 75

References 77

Subject Index 103

1 Introduction

Cells are continually lost and replaced by the tissues of multicellular organisms due to physiological turnover, disease, and injuries. Cell and tissue replacement that maintains or restores the original tissue structure and function is called regeneration. All organisms regenerate, though the degree of regenerative ability varies among species (Goss 1969). Within individual organisms, regeneration takes place on all levels of biological organization, from the molecular to tissues and organs. Tissues that do not regenerate are repaired by fibrosis, or scarring, which patches damaged areas with non-functional tissue. Fibrosis is due to an inflammatory response that results in a fibroblastic granulation tissue that organizes collagen fibrils into thick bundles instead of the normal reticular pattern (Linares 1992; Clark 1996). Examples of non-regenerating tissues are the dermis of the skin, pancreas, spinal cord and brain, neural retina and lens of the eye, cardiac muscle, lung, and kidney glomerulus. Scarring can also occur in tissues that have the ability to regenerate when damage exceeds their regenerative capacity.

The cost of tissue damage due to degenerative disease and injury is enormous in terms of health care (estimated to exceed U.S. $400 billion in the United States alone), lost economic productivity, diminished quality of life, and premature death. The health care costs of spinal cord injuries alone exceed $8 billion per year and $1.5 million per patient over a lifetime in the United States. Currently, organ transplants and bionic implants are our only means of replacing non-regenerating tissues and organs. Tissue and organ transplantation is limited by donor shortages and the need for immunosuppression. There are limitations to our ability to engineer prosthetic and bionic devices that match our tissues and organs in durability, size, shape, and function. What we really desire is the ability to regenerate the original structure and function of damaged human tissues and organs. The emerging field of regenerative medicine aims to do just that and thus has created great excitement in the world of medical science. Potential regenerative therapies include transplantation of stem cells or their derivatives, implantation of bioartificial tissues synthesized in the laboratory, and the chemical induction of regeneration from our own tissues in vivo. What has been forgotten in the excitement, however, is that regenerative medicine will not become a reality without a fundamental understanding of the biology of regeneration. This understanding is far from complete. Thus, it is appropriate to call this new field of tissue restoration "regenerative *biology* and medicine" to emphasize that understanding the basic biology of regeneration is prerequisite to establishing a regenerative medicine.

Adult vertebrates exhibit four types of regeneration:

1. *Regrowth of cellular parts*, as exemplified by the regeneration of peripheral nerve axons (Yannas 2001).
2. *Compensatory hyperplasia*—the proliferation of cells while they maintain their differentiated functions. The liver is the prime example of regeneration by this means (Michaelopoulos and De Frances 1997).
3. *The activation of reserve adult stem cells (ASCs) that reside in differentiated tissues.* This is the most common mechanism of regeneration in multicellular organisms. Vertebrate tissues that regenerate via reserve ASCs are blood, epithelia, skeletal muscle, bone, olfactory bulb, and acinar cells of the pancreas (Stocum 2001). The liver also contains a population of ASCs that are activated when the liver is damaged beyond the capacity for regeneration via compensatory hyperplasia.
4. *The creation of stem cells by dedifferentiation.* Dedifferentiation is the loss of phenotypic specialization that converts differentiated cells into stem cells that proliferate and differentiate into replacement tissue. Teleost fish and certain species of lizards can regenerate fins and tails, respectively, by dedifferentiation. The divas of dedifferentiation, however, are the anuran tadpoles and larval and adult urodele amphibians, which regenerate a wide variety of structures by this means, including lens, neural retina, intestine, upper and lower jaws, tails, and limbs (Stocum 1995).

The different modes of tissue repair involving cell proliferation and differentiation are diagrammed in Fig. 1.

T.H. Morgan (1901) distinguished two basic modes of regeneration, morphallaxis and epimorphosis, in which growth and patterning of new tissues are related in different ways (Carlson 2003). Morphallaxis is the regeneration of missing parts in the absence of growth by repatterning the remaining tissue into a normally propor-

Fig. 1 Types of tissue repair, resulting in either regeneration (NORMAL TISSUE)or fibrosis (SCAR). Regeneration-competent adult stem cells (R-C STEM) are of two types, resident stem cells and stem cells created by the dedifferentiation of differentiated cells (DIFF CELLS). Both types give rise to normal tissue. Differentiated cells can also give rise to normal tissue by compensatory hyperplasia, as in liver regeneration, but the differentiated cells of most tissues cannot do this and the tissue is repaired by scar. Many regeneration-incompetent tissues harbor stem cells (R-I STEM) that respond to injury by forming scar tissue in vivo, but exhibit their regenerative potential when placed in a permissive in vitro environment

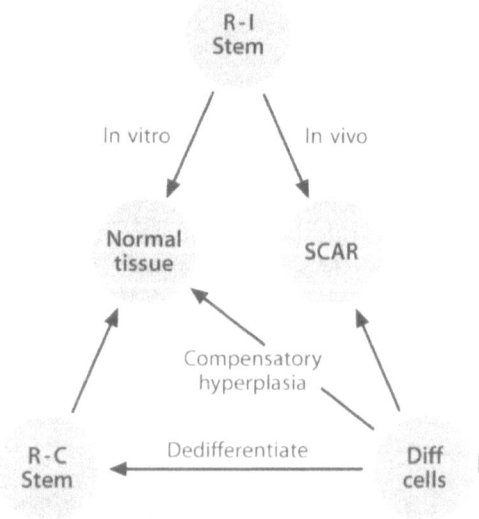

tioned but smaller whole, followed by growth to the original size. It is typical of multicellular animals with relatively simple tissue organizations, such as *Hydra* and planaria. Epimorphosis is regeneration by the proliferation of stem or progenitor cells, followed by their differentiation into the tissues that were lost. Tissue patterning is thus coupled to growth in epimorphic regeneration, whereas it is not in morphallactic regeneration. Epimorphic regeneration is characteristic of annelid worms, crustaceans, fish, lizards, and urodele amphibians (Goss 1969).

The ultimate objective of regenerative biology is to define the permissive and inhibitory factors that determine whether regeneration or fibrosis will take place after injury. This knowledge can then be used to devise therapies to stimulate the regeneration of damaged human tissues that do not regenerate spontaneously, or whose regenerative capacity has been compromised. Currently, we know little about the differences that lead to regeneration as opposed to fibrosis. The failure of a tissue to regenerate could be due either to a lack of regeneration-competent cells, the lack of an environment favorable to regeneration, or both. The fact that regeneration-competent cells can exist as both differentiated or undifferentiated reserve cells suggests that virtually all adult cells have the potential to engage in regeneration, an idea that is compatible with the presence in all cells, with the exception of B and T cells, of a complete genome. This genome can, under natural and experimental circumstances, be reprogrammed to regenerate tissues and appendages (Stocum 2004) or even complete organisms (Byrne et al. 2002; Wilmut et al. 2002). The standard approach to understanding why regeneration fails in mammalian tissues that do not regenerate spontaneously is to lesion these tissues, attempt to identify molecules in the injury environment that are inhibitory to regeneration, and devise ways to neutralize them. While this approach is useful, a more direct strategy is to compare the transcriptomes and proteomes of regenerating vs non-regenerating tissues and define the stimulatory and inhibitory molecules that differentiate regeneration from fibrosis.

In this review, I will first discuss the biology of regeneration and then our progress in translating this biology into therapies that restore tissue structure and function. The focus will be primarily on tissues regenerating by growth of cell parts (axons), compensatory hyperplasia (liver), and activation of resident stem cells. Regeneration by dedifferentiation in urodele amphibians has been recently reviewed in a number of venues and will not be considered here (Stocum 1996, 2004; Brockes 1997; Geraudie and Ferretti 1998; Brockes and Kumar 2002; Nye et al. 2003; Tanaka 2003).

1.1
The Biology of Regeneration

All of the trillions of cells in the adult body, including ASCs, are derived from less than 100 embryonic stem cells (ESCs) of the pre-implantation blastocyst. ESC cultures have been established from the early embryos of fish, birds, and a variety of mammals (Pain et al. 1996; Brook and Gardner 1997; Hong et al. 1998). Human ESC cultures have been established from unused frozen blastocysts produced by in vitro fertilization in assisted reproduction facilities, and from primordial germ cells of 5- to 9-week embryos (Shamblott et al. 1998; Thomson et al. 1998). These cells express markers characteristic of primate pluripotent cells: stage-specific embryonic antigens (SSEA-3 and 4, TRA-1–60 and 81), alkaline phosphatase, and high levels of telomerase. ESCs, whether freshly isolated, or cultured for long periods of time, are pluripotent, capable of giving rise to any of the more than 200 differentiated cell types of the body, including oocytes (Hubner et al. 2003). This pluripotency has been unequivocally demonstrated in vivo by injecting ESCs into host blastocysts, where they make contributions to all tissues to form a chimeric embryo (Smith 2001). In mammals, all the cells of the blastocyst inner cell mass are pluripotent. Prior to implantation, the pluripotent cells become restricted to the epiblast of the embryo. Epiblast cells are self-renewing and pluripotent up to gastrulation, when they begin progressive differentiation into the three germ layers and their derivatives (Smith 2001). The acquisition and maintenance of pluripotency of mouse ESCs requires expression of the leukemia inhibitory factor (LIF)/STAT3-dependent transcription factors OCT4 (Smith 2001), SOX2 (Avilon et al. 2003), and Fox D3 (Hanna et al. 2002). Recently, a LIF/STAT3-independent transcription factor, Nanog, has been discovered that is also essential for mouse ESC pluripotency and self-renewal (Chambers et al. 2003; Mitsui et al. 2003).

Embryonic stem cells give rise to the three germ layers of the embryo—ectoderm, endoderm, and mesoderm—from which differentiated adult tissues emerge. Most of the tissues derived from ectoderm and endoderm are epithelial in nature and represent about 60% of the tissue mass in the adult mammalian body (Slack 2000). These include cornea, retina, lens, epidermis, appendages (hair follicles, sweat glands, sebaceous glands, nails, feathers, scales), kidney tubules, linings of the digestive and respiratory tracts, liver, pancreas, and the central nervous system. The liver turns over very slowly and regenerates by compensatory hyperplasia, but most epithelial tissues turn over rapidly and are continually regenerated via reserve stem cells. Epithelial stem cells and their differentiated progeny are usually located adjacent to one another within the tissue. This spatial relationship has given rise to the concept of "structural-proliferative units" composed of a few stem cells feeding a differentiated compartment (Slack 2000). The remaining 40% of tissues are mesodermal derivatives. These include the musculoskeletal, cardiovascular, blood, and immune systems. With the exception of blood and immune cells, mesodermal derivatives turn over very slowly, but like epithelial tissues they respond to injury with regeneration via reserve stem cells.

1.2
Regeneration of Ectodermal Derivatives

1.2.1
Uninjured Epidermis

The cells of the epidermis are arranged in columnar structural-proliferative units consisting of basal cells at the bottom of each column, two intermediate layers of differentiating keratinocytes, the stratum spinosum and stratum granulosum, and the top layer of flattened keratinized cells, the stratum corneum (Potten 1974). The basal layer of the epidermis is a mixture of epidermal stem cells (EpSCs), transit amplifying cells, and post-mitotic cells committed to undergo terminal differentiation (Potten and Morris 1988). Transit amplifying progeny of the basal cells detach from the basement membrane and migrate upward, undergoing progressive differentiation into the dead, keratinized squames of the stratum corneum. Basal keratinocytes in vitro express three $\beta1$ integrins that mediate adhesion to the basement membrane and regulate the onset of commitment to terminal differentiation and upward migration (Adams and Watt 1990). These are $\alpha_5\beta_1$ (fibronectin receptor), $\alpha_2\beta_1$ (binds to collagen and laminin), and $\alpha_3\beta_1$ (receptor for laminin and epiligrin). Commitment to terminal differentiation and upward migration of basal keratinocytes is associated with the downregulation of the $\beta1$ integrins, resulting in a decreased ability to adhere to the basement membrane (Adams and Watt 1990; Hotchin et al. 1995).

No unique surface antigens or combinations of transcription factors have been identified that specifically define EpSCs. However, there is a population of basal cells that expresses high levels of the transmembrane Notch ligand Delta 1 (Lowell et al. 2000) and 2–3 times the level of $\alpha_2\beta_1$ and $\alpha_3\beta_1$ integrins ("integrin-bright" cells) than surrounding "integrin-dull" cells (Jones and Watt 1993; Jones et al. 1995). Integrin-bright cells constitute about 40% of the cells in the basal layer. This population exhibits a high frequency of differentiation into keratinocytes in vitro, indicating that it is rich in stem cells. EpSCs can be isolated to greater than 90% purity by FACS on the basis of $\beta1$ integrin levels and adhesion assays in vitro. Further evidence for the stemness of integrin-bright cells is that they divide infrequently to give rise to a more profligate transit-amplifying cell population that undergoes 3–5 rounds of division within the basal layer (Jones et al. 1995; Jensen et al. 1999). The transit-amplifying cells spread out over the basement membrane to fill any intermediate spaces (Jensen et al. 1999), commit to keratinocyte differentiation, and move upward toward the epidermal surface, where $\beta1$ integrin expression is downregulated and ultimately extinguished as the cells become terminally differentiated.

EpSCs plated in vitro reconstruct integrin-bright and dull patches of epidermis. This self-organizational tendency suggests that stem cell patterning in the epidermis is an intrinsic property of keratinocytes, rather than depending on environmental cues (Jones et al. 1995). Computer simulations based on the assumption that committed basal keratinocytes migrating upward occupy less crowded regions showed that the epidermal architecture of orthic tetrakaidecahedrons is spontaneously organized from basal cells supplied at random (Honda et al. 1996), consistent with the in vitro evidence for self-organization of epidermal patterning.

1.2.2
Injured Epidermis

The regeneration of epidermis over the fibrin matrix and granulation tissue of an excisional wound is accomplished by lateral migration of basal stem cells. Basal cells begin migrating into the fibrin matrix of the wound within a few hours after injury. The initial signal for migration may be a "free edge" effect, in which a lack of neighbors on one side stimulates the cells to alter their morphology and internal structure to accommodate movement (Woodley 1996). The migrating cells undergo partial dedifferentiation, losing their apical-basal polarity and dissolving the desmosomes that hold them together laterally, as well as the hemidesmosomes that anchor them to the basement membrane. Simultaneously, the epidermal cells form the peripheral actin locomotory apparatus that gives them motility. The cells may move as a sheet, or in "leapfrog" fashion with cells just behind the leading edge moving over the lead cells to establish a new edge. One to two days after the initiation of migration, the epithelial cells next to the wound edge begin proliferating to provide new cells for migration. To facilitate their passage through the fibrin matrix the epidermal cells produce collagenase and plasminogen activator, which activates the collagenase (Woodley 1996). The epidermis continues to migrate during the formation of granulation tissue until the wound is fully covered. Macrophages are thought to play a crucial role in initiating epidermal cell migration via their production of epidermal growth factor (EGF) and transforming growth factor (TGF)-α. These growth factors promote the spreading of epithelial sheets in vitro (Woodley 1996). The epidermal cells themselves are induced to produce these growth factors, which maintain migration through an autocrine mechanism. Activin B is strongly upregulated in the migration and proliferation of keratinocytes, suggesting that this member of the TGF-β family plays an autocrine role in these processes (Hubner et al. 1996).

Epidermal cells upregulate integrin receptors for migration on fibronectin in the wound extracellular matrix (ECM). Tenascin, an anti-adhesive molecule, is synthesized by dermal fibroblasts at the wound edge and is detectable within 24 h after injury (Whitby and Ferguson 1991), its appearance being correlated with the initiation of basal cell migration (Repesh et al. 1982). TGF-β appears in rat skin wounds 12 h after injury (Whitby and Ferguson 1991). TGF-β is strongly upregulated and TGF-α is downregulated in keratinocytes migrating into a full-thickness punch wound in human foreskin in vitro (Kratz et al. 1997). The presence of platelet-derived TGF-β in the wound at 12 h and the appearance of tenascin at 24 h after injury may reflect an inductive effect of TGF-β on tenascin synthesis. A balance of adhesive and de-adhesive forces is thus likely to control the rate of epithelial migration across the fibrin clot and granulation tissue.

Fibroblasts of the granulation tissue produce keratinocyte growth factor (KGF, or FGF-7), which stimulates mitosis in the epidermis, allowing it to increase in thickness (Pierce 1991). Expression of a dominant-negative transgene for KGF receptor in mice reduces the rate of proliferation of keratinocytes at the wound edge, thus substantially delaying re-epithelization of the wound (Werner et al. 1994). The regenerated wound epidermis resynthesizes a new basement membrane as well as type VII collagen anchoring fibrils (Miller and Gay 1992). Epidermal proliferation and thickening are promoted by epiregulin, a member of the EGF growth factor family that enhances the repair of mouse excisional skin wounds (Draper et al. 2003). All members of the EGF family are anchored into the plasma membrane of cells and are cleaved to produce active fragments. The active fragments of epiregulin exert their effects by binding to the tyrosine kinase EGF receptors erbB1 and erbB4 (Komurasaki et al. 1997). Respiratory epithelial cells produce heregulin, which is anchored to the plasma membrane on the apical side of the cells and binds to erb2. When injured, these epithelia regenerate from basal stem cells, as in the epidermis. In the epidermis and respiratory epithelium, tight junctions separate the apical and basal extracellular domains of epithelial cells, preventing any diffusible molecular interactions between the domains. The spatial locations of the epiregulin receptors have not been determined in epidermis, but in respiratory epithelium the erbB2 receptor is on the basolateral surface of the cells. In vitro experiments in which a break was made in the epithelium showed that injury allowed access of heregulin to erbB2, followed by proliferation. The same result was achieved by the removal of calcium, which is required to maintain tight junctions (Vermeer et al. 2003). These results suggest that the injury-promoted access of active fragments of ligands belonging to the EGF family to their receptors on the opposite side of the cell may be a common trigger mechanism for initiation of epithelial stem cell proliferation.

1.2.3
Hair

Mammalian hair is constantly falling out and regenerating from hair follicles, hollow epidermal tubes that project into the dermis. The tip of the follicle forms a cap over the dermal papilla, a condensation of dermal fibroblasts associated with the follicle (Hardy 1992). This cap of follicle cells is called the matrix. Together, the dermal papilla and the matrix constitute the bulb of the hair follicle. The hair shaft is formed by the proliferation and differentiation of the matrix cells upward through the lumen of the follicle. The differentiated hair consists of keratinocytes that form a medullary core surrounded by a cortex. The walls of the hair follicle above the matrix differentiate into inner and outer root sheaths. Two epidermal thickenings are present in the upper third of the outer sheath. The upper thickening forms the sebaceous gland. The lower thickening is called the "bulge" and is the site of attachment of the arrector pili muscle. Dermal fibroblasts surrounding the outer root sheath condense to form the dermal sheath of the follicle.

Hairs are lost and regenerated in three phases (Hardy 1992; Messenger 1993): (1) catagen, the cessation of hair growth and regression of the hair follicle; (2) telogen, or follicular rest; and (3) anagen, the regeneration of a new hair. During catagen, cells in the follicle wall undergo apoptosis and the hair follicle regresses upward. In telogen, the epithelial cells are in a resting state and the hair shaft remains within the shortened follicle. The dermal papilla also shrinks in volume during catagen and telogen, but it is not known if this is due to cell loss or compaction (Matsuzaki and Yoshizato 1998). During anagen, the hair follicle grows back down into the dermis and a new hair shaft and inner root sheath are differentiated from matrix cells. The new hair pushes the old one out of the hair canal as it grows.

The EpSCs responsible for hair follicle growth during anagen reside in the bulge of the external root sheath. Injections of ^{3}H-thymidine or bromodeoxyuridine (BrdU) label a population of bulge cells in both mouse and rat hair follicles that retain the label for much longer periods than other cells of the follicle epithelium, indicating that they cycle very slowly (Cotsarelis et al. 1990; Kobayashi et al. 1993; Lavker et al. 1999; Morris and Potten 1999; Taylor et al. 2000). In rat vibrissae, these cells constitute over 95% of keratinocyte colony-forming cells in vitro. Bulge cells in human scalp hair follicles form clones with proliferative capacity much higher than epithelial cells from other regions of the follicle (Rochat et al. 1994). Direct evidence that the bulge cells give rise to the hair has been obtained from long-term BrdU labeling studies in mice (Taylor et al. 2000). Eight weeks after labeling, all the follicles were in telogen and only bulge cells contained label. By 10 weeks, however, as the follicles entered anagen, many of the matrix cells below the bulge contained label.

Bulge cell progeny also give rise to epidermis outside the hair follicle (Taylor et al. 2000). Mice were given BrdU to label bulge cells of hair follicles in telogen, then were labeled with ³H-thymidine 18 h later. Double-labeled cells were observed above the bulge. The number of these double-labeled cells subsequently decreased while simultaneously increasing in the epidermis surrounding the follicle. Double-labeling of upper follicular cells with BrdU and ³H-thymidine after making a full-thickness wound in the dorsal skin of mice again demonstrated an increase of double-labeled cells in the repairing epidermis and a corresponding decrease in double-labeled cells in the follicular epithelium. Thus, bulge stem cells are the source of both hair and epidermis.

Dermal papilla cells appear to supply signals to epidermal cells for the regeneration of hair follicles (Inamatsu et al. 1998). New hair bulbs regenerate when hair bulbs are removed by amputating the lower third of the follicle (Horne et al. 1986). The upper halves of hair follicles will regenerate new hair bulbs when cultured under the kidney capsule with dermal papillae (Kobayashi and Nishimura 1989) or alone (Matsuzaki et al. 1996). The regeneration of hair bulbs after removal of the lower third or half of the follicle is associated with the regeneration of a new dermal papilla from dermal sheath cells (Matsuzaki and Yoshizato 1998).

Relatively little is known about the molecular signals that regulate the hair growth and regression cycle. Papilla cells express more laminin, fibronectin, and versican than surrounding dermal fibroblasts (Messenger et al. 1991; du Cros et al. 1995). Papilla cells secrete hepatocyte growth factor (HGF) and vascular endothelial growth factor (VEGF) (Shimaoka et al. 1994; Lachgar et al. 1996). HGF stimulates the growth of mouse vibrissal follicles in vitro (Jindo et al. 1994). Hair matrix cells express more TGF-α and platelet-derived growth factor (PDGF)A than basal cells of the epidermis, and epidermal basal cells have receptors for insulin-like growth factor (IGF)-I, whereas follicular epithelial cells do not (Akiyama et al. 1996; Rudman et al. 1997). Clusterin, a sulfated glycoprotein involved in a wide variety of processes such as cell–cell adhesion, apoptosis, and tissue remodeling, is expressed in the inner root sheath during anagen (Seiberg and Marthinuss 1995). *Fgf5* is expressed in the outer root sheath of mouse hair follicles during anagen (Hebert et al. 1994). Mice with null alleles of this gene have abnormally long hair, suggesting that *Fgf5* functions to inhibit hair growth. Recently, it was shown that the transcription factor lef-1 plays a central role in the embryonic development of hair follicles. The activity of *lef-1* is normally suppressed by bone morphogenetic protein (BMP). Follicle formation is associated with the activation of *lef-1* by two signaling pathways. One pathway is indirect, through the inactivation of BMP by noggin, while the other is the direct activation of *lef-1* by Wnt through β-catenin (Millar et al. 1999; Jamora et al. 2003). Activation of *lef-1* results in decreased e-cadherin synthesis, which seems to be requisite for follicle formation.

1.2.4
Nervous System

The nervous system is derived from an epithelial neural tube and has three major divisions: central, peripheral, and autonomic. The central nervous system consists of the brain and spinal cord. It controls the voluntary actions of the body, as well as some involuntary actions, such as reflexes. The retina and optic nerve (cranial nerve

II) are considered part of the central nervous system, since they develop from and project to, the diencephalon of the brain. The peripheral nerves include 31 pairs of mixed (motor and sensory) nerves, as well as cranial nerves I (olfactory) and III–XII, which develop from ectodermal placodes and neural crest. The autonomic system, a complex subset of the peripheral nervous system, controls involuntary activities, such as heart rate, temperature, and the smooth muscle activity of the vascular and digestive systems.

Neural tissue in all three divisions is made up of neurons (nerve cells) and associated glial cells. Neurons consist of a cell body that sends electrical signals over axons and receives signals from the axons of other neurons through shorter dendrites. Neurons are linked at synapses, junctions where axons from one neuron meet the dendrites or cell body of another neuron. All nerve cell bodies and axons are associated with supporting glial cells. In the brain and spinal cord, the major glial cell types are oligodendrocytes, which form insulating myelin, and astrocytes, of which several types reside in both the gray and white matter. Astrocytes are involved in maintaining acid-base balance, degrading and forming synapses, and modulating neuron responses (Travis 1994). The glial cells associated with the optic nerve are primarily astrocytes. In the peripheral nervous system, Schwann cells are the glial counterparts of CNS oligodendrocytes and form the myelin sheath of the nerves. The myelin sheath is surrounded by a basement membrane.

1.2.4.1
Axon Regeneration in Mammals

Axons of both central and peripheral neurons in adult mammals have the intrinsic potential to regenerate after crush or transection, as indicated by the fact that they initiate sprouting after such injuries. In most vertebrates, spinal nerve and olfactory nerve axons are able to regenerate across the lesion and back to their targets. The optic nerve and the ascending and descending tracts of the spinal cord in fish and adult urodele amphibians are also able to regenerate. By contrast, axons of the mammalian central nervous system fail to regenerate further after initial sprouting.

A large body of evidence suggests that whether or not axons regenerate depends in large part on the kind of environment supplied by their associated glial cells. Differences in the ability of peripheral and central glial cell populations to support regeneration have been well documented by experiments in which the regeneration of central axons was promoted by peripheral nerve sheaths grafted into the central nervous system, whereas central nerve sheaths inhibit the regeneration of peripheral axons (Aguayo 1985). These differences appear to reside largely in the adhesion molecules synthesized by glial cells, as well as diffusible signals made by glial and target cells. Glial cells that support regeneration provide most or all of the molecules that are important for neuron survival and axon outgrowth, whereas regeneration fails where glial cells do not make these molecules and/or synthesize molecules inhibitory to regeneration.

Spinal Nerves. The axons of spinal nerves are organized on three levels (Yannas 2001). The fundamental unit is the endoneurial unit, composed of an axon, its associated Schwann cell sheath and basement membrane, and surrounding connective tissue sheath, the endoneurium. Endoneurial units are organized into fascicles, each of which is surrounded by a perineurium. The fascicles are bundled into the nerve trunk, which is encased by the epineurium. The nerve trunks are richly vascularized by epineurial, intrafascicular, and perineurial arteries and arterioles, and the endoneurium contains a network of capillaries.

Mammalian spinal nerves regenerate relatively well, provided that the endoneurial tubes remain intact and in register at the site of injury (Yannas 2001). The ends of the axons on the proximal side of the injury are sealed off and their distal parts degenerate. The proximal axon stumps then regenerate through the endoneurial tube to make new synapses with target skin and muscle (Griffin and Hoffman 1993). Neurons cultured in vitro retract their axons, but on the appropriate substrate will re-extend them. Following a crush injury, the disconnected axons and their myelin layers disintegrate over a period of several days, a process known as Wallerian degeneration. Simultaneously, the Schwann cells associated with the axons dedifferentiate within the basal lamina to form cords of cells called the bands of Bungner. Those dedifferentiated Schwann cells at the cut edge of the distal segment proliferate and migrate to meet Schwann cells proliferating from the edge of the proximal nerve stump, thus bridging the lesion. The end result is an array of endoneurial tubes extending from the proximal stump of the nerve, each filled with a cord of dedifferentiated Schwann cells surrounded by a basal lamina (Yannis 2001). These tubes guide the regenerating axons from the proximal stump of the nerve back to their targets. Regeneration after nerve transection is more problematic. Axons from the proximal stump initiate regeneration, but often form neuromas due to obstruction and misdirection by nerve sheath fibroblasts proliferating in the lesion space (Yannas 2001). In this case, sprouting axons can be directed to their targets by suturing the cut ends of the nerve together through the epineurium. However, the quality of the regeneration is not as good as after a crush injury.

During Wallerian degeneration, myelin is broken down by Schwann cells and macrophages into lipid droplets that contain cholesterol and fatty acids. Electron microscopic, metabolic labeling, and cell surface marker expression studies have shown that the Schwann cells degrade small bits of myelin, but that the bulk of the myelin is phagocytosed and degraded by macrophages that enter the nerve tubes (Goodrum and Bouldin 1996). The cholesterol and free fatty acids produced by the degradation are complexed to apolipoprotein E within the macrophages to form lipoprotein particles. These particles are then released from the macrophages and taken up by Schwann cells via low-density lipoprotein (LDL) receptors, to be reused in myelin synthesis during axonal regeneration.

Axon regeneration has three phases: sprouting, elongation, and maturation (McQuarrie 1983). As Schwann cells dedifferentiate and proliferate, the proximal stumps of the injured axons sprout by the actin-driven formation of growth cones (Sinicropi and McIlwain 1987). As the growth cone advances, it elongates the axon by pulling out a thin daughter cylinder of the proximal axon stump. The elongating part of the axon is stabilized by the polymerization of microtubules behind the growth cone. Growth cone sprouting is inhibited by cytochalasin B, which destabilizes actin microfilaments, and axon elongation is inhibited by colchicine, which destabilizes microtubules. When the growth cone contacts its target, the axon enlarges radially to its mature diameter. Radial maturation is associated with the elaboration of neurofilaments (McQuarrie 1983). Labeling, electron microscope, and immunocytochemical studies have shown that the materials for building new cell membrane and cytoskeleton in the elongating axon are supplied by an upregulation of RNA and protein synthesis in the neuron cell body (Grafstein 1983) and the anterograde transport of the materials. Different kinds of materials are transported distally at different rates. The cytoskeletal subunits are carried distally at the two slowest rates (0.2–2.0 mm/day for neurofilament subunits and 2–6 mm/day for G-actin and tubulins. Other materials are transported as part of the "fast" component (50–400 mm/day) (McQuarrie 1983).

The growth cones of regenerating axons are guided back to their original targets by the endoneurial tubes (Yannas 2001). The Schwann cells within the tubes play a major role in promoting and guiding axon outgrowth (Bunge 1987). The surfaces of the dedifferentiated Schwann cells and the ECM molecules of their surrounding basement membranes provide a substrate on which growth cones of the regenerating axons can extend. In vitro, both are effective in promoting axon outgrowth. As in embryonic development, laminin and fibronectin promote growth cone extension (Rogers et al. 1983; Westerfield 1987). Merosin is the predominant isoform of laminin present in the basal lamina of Schwann cells. Antibodies to merosin inhibit neurite outgrowth and Schwann cell migration in vitro and block the regeneration of sympathetic fibers in vivo (Anton et al. 1994). Fibronectin is present in the basement membrane of the endoneurial tube and is strongly expressed by Schwann cells and fibroblasts. The integrin fibronectin receptor $\alpha_5\beta_1$ is synthesized by both Schwann cells and regenerating axons. Thus, fibronectin may be important both for Schwann cell migration and for neurite extension on Schwann cells during regeneration (Lefcourt et al. 1992).

The results of experiments in which neurons are cultured in the presence or absence of target tissues indicate that, as in the embryonic development of the peripheral nervous system, target tissues produce diffusible growth and neurotrophic factors essential for axon survival (Kuffler 1987; Birling and Price 1995; Garrity and Zipursky 1995). IGF-I and -II, PDGF, and FGF-1 and -2 promote axon extension. Nerve growth factor (NGF) synthesized by target tissues and transported retrogradely is important to the survival of sensory and sympathetic neurons. The sciatic nerve exhibits low levels of NGF mRNA synthesis. Following transection, all the non-neuronal cells of the distal part of the proximal nerve stump synthesize NGF mRNA, but not NGF receptor, whereas the reverse is true for the regenerating axons. The axons take up NGF and transport it retrogradely. NGF mRNA synthesis is also upregulated in Schwann cells of the endoneurial tube distal to the lesion. The significance of this upregulation is unknown, but may indicate that NGF has a mitotic or trophic effect on Schwann cells (Heumann 1987). Glial growth factor (GGF) is a mitogen for Schwann cells, while the neurotrophic factors brain-derived neurotrophic factor (BDNF), ciliary neurotrophic factor (CNTF), GDNF, NT-3 and NT-4/5 promote survival of regenerating motor neurons (Birling and Price 1995).

Olfactory Nerve. The olfactory nerve is a cranial nerve arising from nerve cell bodies in the nasal epithelium that are specialized for discriminating between different odorants. Olfactory receptor neurons have a life span of 4–8 months, depending on species, and are replaced continuously throughout the life of the animal by neurons regenerated from stem cells located in the deepest layer of the olfactory epithelium

(Schwob 2001). If the olfactory nerve is transected, the receptor neurons degenerate and are replaced in the same way (Ramon-Cueto and Valverde 1995). Either way, the axons of the replacement neurons must extend into the olfactory bulb to synapse with neurons there.

The ability of olfactory axons to extend into the olfactory bulb, in contrast to the inability of other CNS mammalian axons to regenerate, is due to their association with olfactory ensheathing glial cells derived from the olfactory placode during development (Mendoza et al. 1982; Chuah and Au 1991). These cells are distinct from other mammalian glial cell populations in terms of their ultrastructure, cytoskeleton, and cell surface molecules expressed, but express the same kinds of survival and growth-promoting factors as other glial cells that support axon regeneration (Ramon-Cueto and Valverde 1995).

No other mammalian cranial nerves are able to regenerate, but the acoustic nerve (cranial nerve VIII) of frogs regenerates well, restoring the specificity of central connectivity that characterizes the auditory system in normal animals (Zakon and Capranica 1981). It would be interesting to know whether this regeneration, too, is dependent on an associated glial cell population and what the characteristics of these cells are.

Spinal Cord. In mammals, spinal cord injury that destroys substantial numbers of axons and neurons causes sensory deprivation and paralysis below the level of injury, followed by muscle atrophy and spasticity. Cervical injuries result in disturbances in autonomic functions such as regulation of blood pressure, heart rate, and temperature. The spinal cord of mammals contains NSCs in the ependyma and subventricular zone (Weiss et al. 1996; Kehl 1997; Doetsch et al. 1999; Johansson et al. 1999), but these cells respond to injury by the formation of glial scar.

If the spinal cord is transected in a way that minimizes displacement of the ends of the cord and infiltration of fibroblasts, it regenerates accompanied by a high degree of functional recovery (Seltzer et al. 2002). Spinal cord axons do not regenerate after crush injury or transection through the pia mater because the molecules necessary to support the extension of growth cones are not expressed by associated glial cells. Instead, these cells contribute to an environment that is inimical to regeneration. Trauma to the spinal cord induces a cascade of events that lead to glial scar formation (Barron 1983; Steward et al. 1999; Chernoff et al. 2003).

First, blood flow is interrupted, depriving the injured area of oxygen and glucose. Excess plasma leaking from damaged vessels causes the cord to swell, further compressing the tissue and killing many neurons and glial cells outright.

Second, undamaged neurons become overexcited, releasing excess amounts of the neurotransmitter, glutamate. Glutamate kills undamaged neurons by opening membrane channels that allow the influx of toxic amounts of calcium, a phenomenon called "glutamate toxicity." A further source of toxicity is the production of free radicals by lipid peroxidation.

Third, the insulating myelin from both dead and surviving axons breaks down. Within a few hours after injury, the breakdown products of myelin and dead cells begin to spread the initial damage to neighboring uninjured regions, a process that can go on for weeks, yielding secondary damage even greater than the primary damage caused by the trauma and causing many intact neurons to undergo apoptosis.

Fourth, oligodendrocytes secrete individual myelin proteins and other molecules into the lesion that have been shown to inhibit axon growth in vitro (Kapfhammer and Schwab 1992; Filbin 2000). One such protein, dubbed "Nogo," is secreted into the lesion at high levels by oligodendrocytes (Chen et al. 2000; Grandpre et al. 2000; Prinjha et al. 2000). Another is myelin-associated glycoprotein (MAG) (Filbin 2000). Chondroitin sulfate proteoglycans (CSPGs) synthesized by oligodendrocytes also inhibit the growth of axons in culture (Niederost et al. 1999) and there are undoubtedly many more inhibitory proteins awaiting identification.

Fifth, as the damage spreads, an inflammatory response is mounted. Platelet growth factors attract neutrophils and macrophages into the lesion from the blood to ingest bacteria and debris. The cleanup of debris continues over the whole area of damage, creating large cavities in the cord tissue. Astrocytes and fibroblasts from the meninges covering the cord proliferate around the periphery of the cavities to form a "glial scar." The astrocytes of the scar produce the same types of inhibitory CSPGs as oligodendrocytes (Filbin 2000). The glial scar thus constitutes both a physiological and mechanical barrier to axon regeneration.

1.2.4.2
Regeneration of Amphibian and Avian Optic Nerves

The optic nerve is a cranial nerve that develops from the retinal evagination of the diencephalon of the brain. It is formed by the fasciculation of axons emanating from retinal ganglion cells, which constitute the inner layer of the neural retina. The optic nerve does not regenerate in mammals, but does so in fish, frog tadpoles, and adult urodeles. In these animals, severing the optic nerve or cutting off the blood supply of the retina results in the complete degeneration of retinal neurons and their axons. Like the olfactory nerve, retinal neurons are regenerated by epithelial stem cells; the optic nerve is then re-formed by axons extending from the newly formed ganglion cells. In this case, the stem cells are created by dedifferentiation of pigmented epithelial retina cells which proliferate and transdifferentiate into all of the cell types of the neural retina. New retinal ganglion cells then extend their axons into the brain and restore normal retinal–tectal connections, with full visual recovery (Sperry 1944; Gaze 1959; Attardi and Sperry 1963).

The axons of regenerating amphibian retinal ganglion cells are dependent on astrocytes associated with the optic nerve to find their way back to their targets in the optic tectum. The astrocytes in the degenerated central stump of the optic nerve hypertrophy and form a longitudinal band within a basement membrane made by the pia mater, analogous to the Schwann cell bands in regenerating peripheral nerves (Gaze and Grant 1978). In contrast to the mammalian CNS, where astrocytic scar impedes axon extension, the astrocytes of the fish and amphibian optic tract promote it (Reier and Webster 1974; Turner and Singer 1974; Stensaas and Feringa 1977; Reier 1979; Scott and Foote 1981). The growth cones of the regenerating axons associate preferentially with endfeet of the astrocytes, which project toward the pia, so all the regenerating axons are found just under the pia (Gaze and Grant 1978; Bohn et al. 1982). Eliminating the astrocyte band by resection of a segment of the optic nerve results in a decrease in the number of axons crossing the lesion into the central nerve stump and an increase in the number of axons deviated into inappropriate locations (Bohn et al. 1982). The guidance substrate preferred by regenerating optic nerve axons of stage 47–50 *Xenopus* tadpoles is laminin, followed by collagen I>polylysine=polyornithine>fibronectin (Grant and Tseng 1986).

Different cell populations of the avian retina can be selectively destroyed by different agents (Fischer et al. 1998; Fischer and Reh 2002). *N*-methyl-D-aspartate (NMDA) destroys amacrine and bipolar neurons, kainite eliminates these neurons plus ganglion cells, and colchicine destroys only ganglion cells. In response to NMDA-induced excitotoxic damage or destruction by kainate, numerous Muller glia in the retina dedifferentiate, proliferate, and express proteins characteristic of embryonic retinal progenitors (Fischer and Reh 2001). Most of the cells remain undifferentiated, but a few differentiate into Muller glia cells and amacrine or bipolar neurons (Fischer and Reh 2002). Destruction of ganglion cells by colchicine results in the regeneration of ganglion cells that is stimulated by a combination of insulin and FGF-2 (Fischer and Reh 2002; Fischer et al. 2002). Whether the optic nerve regenerates to make functional connections with visual recovery in these experiments was not reported.

1.2.4.3
Axon Regeneration in the Amphibian Spinal Cord

Larval and adult urodeles regenerate the axons of ascending and descending nerve tracts after transection or ablation of the spinal cord at the level of the trunk, with recovery of function (Piatt 1955; Butler and Ward 1965; Egar and Singer 1972; Nordlander and Singer 1978; Chernoff et al. 2003; Ferretti et al. 2003). The ependymal (NSC) layer of the cord plays a central role in this process (Holder and Clarke 1988; Ferretti et al. 2003; Chernoff et al. 2003). The ependymal cells span the radius of the cord, branching within the white matter to terminate in expanded endfeet that form the glia limitans under the pia mater (Holder et al. 1990). These cells resemble the radial glial cells of the embryonic spinal cord of birds and mammals, which can regenerate (Clarke and Ferretti 1998; Ferretti et al. 2003). However, they are different from the radial glial cells of birds and mammals in that they express GFAP, but not nestin or the intermediate cytoskeletal filament vimentin. They also express the epithelial keratins 8 and 18, which are not expressed in the ependymal cells of mammals, but are expressed in the ependymal cells of fish and lampreys, which can also regenerate the spinal cord (Holder et al. 1990; O'Hara et al. 1992; Bodega et al. 1995; Clarke and Ferretti 1998; Chernoff et al. 2003). These features are sometimes cited as possible correlates of regenerative potential. However, this is unlikely given that adult frog ependymal cells share these features but do not regenerate the spinal cord (Chernoff et al. 2003).

The ependymal response to injury appears to provide an environment that protects axon stumps from degenerating and promotes their growth to make new synaptic connections that allow the recovery of function. After transection, the ependymal cells on either side of the lesion undergo an epithelial to mesenchymal transformation in which they suppress production of laminin, GFAP, and epithelial cytokeratins and express fibronectin and vimentin (O'Hara et al. 1992; Chernoff et al. 1998). The mesenchymal cells proliferate to form blastemas that subsequently fuse to bridge the gap between the cut ends of the cord. The proliferating mesenchyme then undergoes a mesenchymal to epithelial transformation to become an ependyma again. The ependymal cells become radially arranged and their endfeet form channels through which axons regenerate, similar to the regenerating axons of the optic nerve (O'Hara et al. 1992; Chernoff et al. 2003). In newts, the regenerated cord is thinner than the intact cord, there are fewer axons and not all the connections made by regenerating axons are correct after weeks or months (Stensaas 1983; Davis et al. 1989), but there is functional recovery of swimming (Davis et al. 1990). In axolotls, innervation from the brain reached control levels in long-term studies of up to 23 months (Clarke et al. 1988).

How the ependymal cells protect and encourage extension of regenerating axons is a major research issue. One possibility is that they may be able to buffer the effect of glutamate excitotoxicity by the uptake and sequestration of calcium, which in turn might trigger epithelial to mesenchymal transformation through second messenger pathways (Chernoff et al. 2003). Retinoic acid (RA) promotes axolotl neurite outgrowth in vitro. The ependymal cells can take up retinol, convert it to retinaldehyde and then to RA. The secreted RA is then taken up by neurons (Hunter et al. 1991). Endogenous retinol and RA have been detected in the urodele spinal cord, implying that RA might be a crucial molecule for axon extension in vivo. Interestingly, the motor neurons of patients who died from motor neuron disease have been found to be deficient in their production of RALDH2, the enzyme that converts retinaldehyde to RA (Corcoran et al. 2002). Most or all of the other survival and guidance factors that have been shown to be important for peripheral nerve regeneration would presumably also be involved in urodele spinal cord regeneration, but this question has not yet been addressed.

In addition, molecules inhibitory to regeneration in mammals appear not to be present or to be removed after spinal cord injury in amphibians. The Nogo protein is not present in the regenerating cord of fish and larval *Xenopus*, although other inhibitory molecules such as MAG, cell surface proteins (CSPs), and tenascin-R are present (Wanner et al. 1995; Lang et al. 1995; Becker at al 1999). Tenascin R and MAG are rapidly removed after injury in spinal cord and newt optic nerve (Becker et al. 1999). In metamorphosing *Xenopus*, spinal cord myelin becomes non-permissive to axon outgrowth and reacts with the anti-Nogo antibody (Lang et al. 1995). The presence and disposition of these molecules have not yet been studied in the regenerating urodele cord.

Delta-Notch is an important signaling pathway for the maintenance of stemness (Go et al. 1998). Axolotl and *Xenopus* ependymal NSCs have been identified that express nrp-1, an RNA-binding protein that maintains the activity of Notch (Chernoff et al. 2003). The factors involved in the transformation of ependymal cells to mesenchyme and back again are incompletely known. Matrix metalloproteinases (MMPs) may play a role in this transformation by digesting ependymal ECM. MMP activity is not detectable in the uninjured cord and tissue inhibitor of metalloproteinase (TIMP)-1 is expressed. The ependymal mesenchyme of the injured cord expresses MMP-1, 2, and 9 and TIMP-1 expression disappears (Chernoff et al. 2003). TGF-β1 and PDGF in combination cause ependymal mesenchyme blastemas to break apart and migrate on the dish as cords, suggesting that these growth factors play a role in modulating the organization of the cells (O'Hara and Chernoff 1992, 1994). Experiments in vitro on ependymal cells from regenerating axolotl spinal cord suggest that migration and proliferation of mesenchymal cells from cultured ependymal blaste-mas in vitro is dependent on EGF, and is inhibited by TGF-β1 (Chernoff et al. 2003).

1.2.4.4
Neurogenesis in the Regenerating Amphibian Tail

Histological and labeling studies have shown that new spinal cord neurons are pro-duced relatively frequently in uninjured juvenile axolotls up to 6–7 months of age, but infrequently in older animals (Holder et al. 1991). Likewise, few or no new neu-rons are generated during the regeneration of axons after transection or ablation of trunk cord in adult salamanders or newts (Butler and Ward 1965; Nordlander and Singer 1978; Davis et al. 1989). By contrast, urodele amphibians can regenerate the spinal cord, including new neurons, after amputation of the tail (Clarke and Ferretti 1998). Urodele spinal cord contains nine groups of neurons, but the fate of each type during either gap replacement regeneration or caudal regeneration from the ampu-tated tail has not been studied (Chernoff et al. 2003).

Tail regeneration in larval urodeles is accomplished by the dedifferentiation of cartilage, muscle, and dermal fibroblasts to mesenchymal stem cells (Echeverri et al. 2001). These cells form a bud of proliferating progenitor cells, the blastema, over the cut surface of the tail. The mesenchymal cells differentiate into new cartilage, muscle, and dermis of the regenerated tail. A separate tube of dividing ependymal, or radial glial, NSCs, extends from the cut end of the spinal cord into the blastema. These ependymal cells maintain their epithelial organization during tail regeneration, as opposed to the transformation into mesenchyme characteristic of gap replacement regeneration. The central canal of the spinal cord is sealed off at the amputation plane by the formation of a terminal vesicle of non-proliferating ependymal cells. Ependymal cells proximal to the vesicle proliferate to extend the ependymal tube (Holtzer 1952; Egar and Singer 1972; Nordlander and Singer 1978). Nestin and vi-mentin are expressed by the proliferating ependymal NSCs (Ferretti 2000). ECM and cell adhesion molecules expressed during spinal cord development are re-expressed by ependymal cells during regeneration (Caubit et al. 1994; Clarke and Ferretti 1998). The distribution of tenascin is like that in developing animals (Caubit 1994). The polysialylated embryonic form of N-CAM (PSA-N-CAM) is expressed at a low level in the ependymal cells of uninjured tail cord, but is strongly upregulated in the regenerating cord (Caubit et al. 1993). Studies in which ependymal cells in the regenerating tail were pulse-labeled with BrdU have shown, using confocal microsco-py, that they give rise to new glia (GFAP[+]) and neurons (NSE[+]) in vivo and in vitro (Benraiss et al. 1999).

The regenerating ependymal tube resembles the neural tube of the embryo. As the tube grows distally, the proliferating cells closest to the amputation plane become ra-dially arranged and extend endfeet to form a glia limitans. The endfeet overlap to form channels that guide the outgrowth of regenerating axons from nerve cell bodies above the level of transection, just as they do after spinal cord transection or ablation (Simpson 1983; Clarke and Ferretti 1998). Some of the proliferating ependymal cells differentiate into new neurons and glia so that the regenerated cord contains new sensory ganglia, motor neurons, and interneurons (Piatt 1955; Geraudie et al. 1988;

Holder et al. 1991; Echeverri and Tanaka 2002). The ependymal NSCs also regenerate neural crest derivatives, such as Schwann cells, fin mesenchyme, and pigment cells (Benraiss et al. 1991). In lizards and adult newts, however, no new neurons are formed, and the spinal cord of the regenerated tail consists only of axons (Duffy et al. 1992; Stensaas 1983; Benraiss 1997).

There is little information about how ependymal NSC proliferation during tail regeneration is regulated. FGF-2 may play a central role. It is not expressed in the ependymal cells of the uninjured cord, though it is expressed in a subset of neurons. In the proliferating ependymal cells of the amputated *Pleurodeles* tail, there is strong expression of FGF-2 and FGF receptors that gradually decreases as differentiation of the new tail progresses until it can no longer be detected in the ependyma. Exogenous FGF-2 injected into the regenerating tail increases the number of proliferating ependymal cells (Zhang et al. 2000). These observations suggest that FGF-2 is important for ependymal cell proliferation after tail amputation. This idea is further supported by the fact that FGF-2 is an important factor in the proliferation of blastema cells during urodele limb regeneration (see later) and it fits with the fact that FGF-2 can enhance the regeneration of pyramidal neurons of the rat hippocampus (Nakatomi et al. 2002; see later). Interestingly, FGF-2 expression is upregulated after CNS injury in mammals by fibrous astrocytes (Fawcett and Asher 1999) and keeps NSCs dividing and undifferentiated in vitro (Svendsen and Caldwell 2000; Temple 2001). This suggests that in both the spinal cord of mammals and urodele amphibians, FGF-2 is a survival and proliferation factor, but that in mammals these functions are negated by factors inhibitory to regeneration in mammals that are not present in urodeles.

Little is known about the signaling molecules and transcription factors that control patterning of neural differentiation and synaptogenesis in the regenerating tail, during gap regeneration, or in the injured mammalian CNS. *Wnt-10*, a member of the *Wnt* family of glycoprotein signaling molecules, is expressed in the uninjured cord of the adult newt *Pleurodeles waltlii* (but not in the cord of other species) and is upregulated weakly in the regenerating *Pleurodeles* cord (Caubit et al. 1997a,b). The *Wnt* proteins play crucial roles in the development of many body structures, particularly the nervous system, and it is possible that maintenance of *Wnt-10* expression in the adult *Pleurodeles* cord and regenerative ability are causally related (Clarke and Ferretti 1998). How the outgrowth and synaptic patterns of axons and dendrites are regenerated is not known. *PwDlx-3*, a homeobox transcription factor related to the *distal-less* gene, which is essential for appendage development in *Drosophila* (Cohen and Jurgens 1989), is upregulated in the ventrolateral region of the cord in the regenerating *Pleurodeles* tail (Nicolas et al. 1996). This is the region from which cells migrate to form the spinal ganglia of the regenerated cord (Geraudie and Singer 1988). *PwDlx-3* is Expression of the gene is not detected in the mammalian central nervous system; thus it has been suggested that in urodeles, its expression might be related to regenerative capacity (Nicolas 1996). Another important gene involved in the patterning of motor neurons in the developing spinal cord is sonic hedgehog (*shh*) (Altmann and Brivanlou 2001), but patterns of *shh* expression have not yet been reported in regenerating amphibian spinal cord.

1.2.4.5
Neurogenesis in the Adult Mammalian Brain

Forty-five years ago the dogma was that, with the exception of the olfactory bulb, neurogenesis ceased after birth in mammals. Then, in the early 1960s, ^{3}H-thymidine labeling studies in rats revealed cells actively synthesizing DNA in the hippocampus, suggesting that physiological regeneration is taking place in this region of the brain (Messier et al. 1958; Smart 1961; Altman 1962, 1963). These studies were essentially ignored until Nottebohm and colleagues showed in the 1980s that there is a tremendous temporary increase in the number of neurons in the vocal control nuclei of male canary brains in the spring of each year. These neurons are recruited into song-learning circuits and die when the mating season is over (Goldman and Nottebohm 1983; Paton and Nottebohm 1984). These studies showed conclusively that there is neuron turnover and replacement in the adult vertebrate brain.

Over the last decade or so, neural stem cells (NSCs) that generate new neurons and glia in the adult mammalian brain have been identified by labeling the DNA of NSCs with an agent such as BrdU and showing that the labeled cells express first stem cell markers (nestin), then markers for immature neurons (doublecortin, Dc) and finally mature neurons (neuron-specific enolase, NSE; γ-aminobutyric acid, GABA; substance P; class III β-tubulin, β-TuIII; or neuron nuclear transcription factor, NeuN). It is also important to show that the BrdU-labeled cells exhibit the morphology and electrophysiology of functional neurons. NSCs purified by FACS have been labeled in vitro with BrdU, or taken from animals transgenic for LacZ or green fluorescent protein (GFP) and subjected to clonal analysis in vitro to test for differentiation into neurons and glia and for self-renewal.

Reynolds and Weiss (1992) showed conclusively that the striatum, a region of the brain that does not normally regenerate, nevertheless harbors NSCs. They enzymatically dissociated adult mouse striatal tissue and cultured the cells in serum-free medium in the presence of EGF, a growth factor known to be important in neural development and whose receptor is expressed in the adult mouse CNS. Most of the cells died, but a few adhered to the culture dish, proliferated, and formed floating neurospheres that were reactive to anti-nestin antibodies. When plated in dishes coated with the adhesive molecule poly-L-ornithine, cells migrated from the neurospheres and proliferated. Sixty-one percent of the migrated cells and the cells within the neurosphere differentiated as astrocytes and were immunoreactive for glial fibrillary acidic protein (GFAP), while 39% were reactive for neuron-specific enolase (NSE) and the neurotransmitters GABA and substance P. Other cells were able to make new neurospheres, indicating that the NSCs had self-renewed.

Similar in vitro studies have led to the conclusion that NSCs reside in several regions of the adult CNS. However, in vivo they actively produce new neurons in only two regions, the subgranular zone of the hippocampus, where they give rise to granule cells, and the ependyma and subventricular zone of the lateral ventricles of the telencephalon, where they give rise to olfactory bulb neurons. Proliferating NSCs observed in other regions of the brain normally give rise only to new glial cells (Gage 2000; Temple 2001).

Olfactory Bulb. Olfactory bulb neurons are constantly turned over and replaced by NSCs in the walls of the lateral ventricles (Schwob 2001). BrdU-incorporation experiments on adult mice have shown that the ependymal cells of the lateral ventricles are labeled first and are therefore the actual stem cells (Johansson et al. 1999; Rietze et al. 2001). The ependymal cells divide asymmetrically to self-renew and give rise to a transit-amplifying population in the subventricular zone (Doetsch et al. 1999; Rietze et al. 2001). These subependymal cells migrate into the olfactory bulb, where they differentiate into new neurons and glia (Lois and Alvarez-Buylla 1994).

NSCs in the ependyma of the lateral ventricles have been purified by FACS fractionation followed by testing the fractions for their ability to form neurospheres under clonal conditions (Rietze et al. 2001). This procedure revealed a discrete population of cells that was nestin[+], had low binding affinity for peanut agglutinin (PNA) and heat-stable antigen (HSA), and did not express differentiated neuronal or glial cell markers. Neurospheres derived clonally from this population differentiated into astrocytes (GFAP[+]), oligodendrocytes (O4[+]), and neurons (beta-tubulin type III[+]). Further evidence that the PNA[lo] HSA[lo] cells were NSCs was a sixfold reduction in the percentage of these cells in the mouse mutant *querkopf*, which has a greatly reduced number of olfactory neurons (Rietze et al. 2001).

Hippocampus. Neurogenesis appears to be a routine process in the adult mammalian hippocampus, an area of the brain that is crucial for cognitive activities, learning and memory. Proliferating NSCs have been found in the hippocampus of mice and rats (Rao 1999; Gage 2000; Momma et al. 2000; Rietze et al. 2000, 2001; Geuna et al. 2001), marmoset monkeys (Gould et al. 1998), and human patients given BrdU as part of a cancer study (Eriksson et al. 1998). Human dentate gyrus cells purified by FACS and transfected with the GFP gene under the control of the nestin enhancer or the Tα1 tubulin promoter proliferated in vitro and differentiated morphologically, antigenically, and electrophysiologically as neurons (Roy et al. 2000).

The ongoing neurogenesis in the hippocampus suggests that NSCs are crucial for the maintenance or formation of memory and for learning new information and tasks. Several experimental results support this idea. Studies in vivo strongly suggest that the number of new neurons born in the hippocampus of mice is influenced by both physical and cognitive activity. Running on a treadmill, or placement in an enriched environment (more mice per cage to increase social interactions, mouse toys and treats, and rearrangeable sets of tunnels) increased stem cell proliferation and neurogenesis in the dentate gyrus of mice above the level of controls (Kempermann et al. 1998; van Praag et al. 1999a,b). Enriched-environment mice also learned a maze faster than controls. Control mice exhibited a decline in the number of BrdU-labeled new neurons in the hippocampus with age that was correlated with a reduction in the speed at which the maze was learned. The reduction in number of BrdU-labeled differentiated neurons in aged animals was reduced by more than half when they were placed in an enriched environment (Kempermann et al. 1998).

Other studies have shown that the new neurons born in the hippocampus behave functionally as neurons and are integrated into hippocampal neural circuitry. Van Praag et al. (2002) transfected proliferating cells in the mouse hippocampus with a retroviral GFP transgene. GFP-labeled cells differentiated into mature neurons that received synaptic input, as shown by ultrastructural analysis and immunostaining with synaptophysin. Electrophysiological recordings indicated that the neurons functioned normally and received normal input from the perforant pathway, the main excitatory input to dentate granule cells. Shors et al. (2001) used behavioral studies to demonstrate integration into the normal hippocampal circuitry. Rats were "trace conditioned" to associate a neutral tone with delayed noxious eyelid stimulation. Neurogenesis in the hippocampus was then inhibited by the DNA methylating agent methylazoxymethane (MAM), which kills proliferating cells. The number of new neurons generated was reduced by 80%, which in turn reduced the frequency of conditioned responses by 50%.

Neocortex. Whether or not new neurogenesis routinely occurs in the neocortex is controversial. Gould et al. (1999) injected macaque monkeys with BrdU and assessed incorporation into cells in the prefrontal, posterior, parietal, and inferior temporal cortex at 2 h and at 1–2 weeks post-injection. At 1–2 weeks, labeled cells were observed to stream through the subcortical white matter into the cortex. Most of the labeled neurons within the cortex reacted for markers associated with mature neurons (MAP-2, NeuN), whereas the cells in the streams outside the cortex reacted with a marker for immature neurons, TOAD-64. These observations suggested that NSCs in the walls of the lateral ventricles replenish cortical neurons. Injection of dyes into cortical areas known to be projection targets of the newly differentiated neurons resulted in their retrograde filling, indicating that they had extended axons that became part of the cortical circuitry.

Another study of BrdU-injected adult macaques also revealed labeled cells that moved to the cortex. These cells failed to express markers of mature neurons but did express GFAP, indicating that they differentiated as glia (Kornack and Rakic 2001). Detailed examination of sections with confocal optics suggested that the close apposition of labeled glial cells and neurons could give the illusion of BrdU-labeled cells that also expressed neuronal markers. Magavi et al. (2002) reached a similar conclusion with regard to the mouse cortex. They, too, observed BrdU-labeled cells in the walls of the lateral ventricles that entered the cortex, but these cells differentiated as glia, remained undifferentiated, or died.

1.2.4.6
Injury-Induced Neurogenesis

There is evidence that injured mammalian brain tissue can respond to injury by increased neurogenesis. Although Magavi et al. (2000) found no evidence of ongoing neurogenesis in the mouse cortex, they did observe neurogenesis after destruction of a subset of pyramidal neurons in the lower layer (VI) of the cortex that project to the thalamus. Thalamic neurons were injected with chromophore-conjugated chlorin e_6 nanospheres, which were retrogradely transported to layer VI cortical neurons. The layer VI neurons were then killed by activating the chromophore with laser light at a wavelength of 674 nm. Subsequent proliferation of BrdU-labeled lateral ventricle cells was no greater than in uninjured animals, but cells labeled with BrdU and expressing doublecortin, a marker of immature neurons, were observed along a path to the cortex. One to two percent of the BrdU-labeled cells within the cortex expressed the mature neuron marker NeuN. Retrograde labeling showed that the regenerated neurons were pyramidal neurons that projected to the thalamus; i.e., the neurons that were destroyed were selectively regenerated and functionally integrated into the normal circuitry of the brain. Other studies have shown increased production of granule cells in the dentate gyrus of the hippocampus in rats and gerbils after neuronal degeneration induced by focal or global ischemia (Gould and Tanapat 1997; Liu et al. 1998; Jin et al. 2001). Thus it would appear that injury signals are sufficient to increase differentiation of NSCs into neurons, but only a small fraction of the degenerated neurons are replaced, not enough for functional recovery.

Ischemic injury alone induces only a small increase in the frequency of differentiation of hippocampal NSCs into pyramidal neurons, but this increase can be markedly enhanced by exogenous FGF-2 and EGF (Nakatomi et al. 2002). Intraventricular injection of FGF-2 and EGF boosted the number of regenerated neurons to 40% of the number that were lost. The neurons were functionally integrated into the normal hippocampal circuitry as determined by microscopy, the electrophysiological properties of synapses, and the performance of the rats on behavioral tasks. Neonatal astroglia induce the differentiation of adult hippocampal NSCs into neurons in vitro, whereas the effect of adult astrocytes is only half that of neonatal cells (Song et al. 2002). Furthermore, the effect seems specific to hippocampal astrocytes, since spinal cord astrocytes do not support hippocampal neurogenesis. These results suggest that the low level of increase in neurogenesis in the injured hippocampus is due to deficient FGF-2 and EGF signaling from astrocytes to NSCs.

1.3
Regeneration of Endodermal Derivatives

1.3.1
Liver

The liver is the classic example of regeneration by compensatory hyperplasia. The hepatocytes of the liver function in both an endocrine and exocrine secretory capacity, reflected in their tremendous amount of mitochondria, rough endoplasmic reticulum, and Golgi stacks. The endocrine activity involves the conversion of glycogen to glucose and the secretion of a large number of serum factors, including albumin, prothrombin, fibrinogen, and the protein component of lipoproteins. The exocrine secretion of hepatocytes is bile, which contains both waste products, such as bilirubin, and bile salts required for intestinal absorption. Hepatocytes also have a major role in carbohydrate, ammonia, and triglyceride metabolism, as well as detoxifying metabolic byproducts and toxic substances.

The structural organization of the liver reflects these secretory and metabolic functions. The hepatocytes, which constitute 80% of the liver, are arranged as trabeculae two cells thick (Ham and Cormack 1978). Spaces between the hepatocytes of the trabeculae constitute biliary canaliculi, which convey bile to the bile ducts. The trabeculae are separated by vascular sinusoids lined by fenestrated endothelial cells and macrophages called Kupffer cells. Between the fenestrated endothelium of the sinusoids and the hepatic trabeculae is the space of Disse. The fenestrated endothelium and the space of Disse provide the hepatocytes with maximum exposure to hepatic blood flow. The hepatocytes have numerous microvilli projecting from their surface into the space of Disse, providing them with a large surface area for absorption of molecules from the blood. Interspersed between hepatocytes and also abutting on the space of Disse are Ito cells, lipocytes that store vitamin A.

The hepatic trabeculae in mammals such as the rat and pig are organized into lobules. The lobules are defined by "portal areas" connected by connective tissue septa to roughly form a hexagon. Hepatic trabeculae radiate from the periphery of the lobule to a central vein. In humans, the connective tissue septae outlining lobules are not present, but the arrangement of the portal areas is the same and lobules can be "seen" by drawing imaginary lines between the portal areas. The portal areas themselves consist of a branch of the portal vein, hepatic artery, and bile duct, together with a lymphatic (Ham and Cormack 1978).

The ECM of the liver is concentrated mainly in its outer connective tissue capsule, blood vessels, and bile ducts, with only small amounts associated with hepatocytes. Collagens I, II, V, VI, VII, fibronectin, and tenascin are found in the outer capsule and blood vessels. Blood vessels and bile ducts all have basement membranes containing laminin, entactin, collagen IV, and perlecan. The sinusoids, however, lack basement membrane. Fibronectin is abundant in the space of Disse, and free collagen IV is also found there (Martinez-Hernandez and Amenta 1995).

The mammalian liver has a very slow turnover time, with the average life span of hepatocytes estimated at ~200–300 days (Bucher and Malt 1971). Cell marking studies have shown that during normal liver turnover, hepatocytes are replaced by compensatory hyperplasia, not by stem cells (Ponder 1996; Grompe and Feingold 2001). All vertebrates exhibit the capacity to regenerate the liver following tissue loss due to trauma, chemical insults, and viral infections. The most widely used, and best studied, model for liver regeneration is partial hepatectomy (PH) in the rat, in which two-thirds of the liver is surgically removed (Higgins and Anderson 1931). The rat liver has four lobes, two large and two small. Excision of the two large lobes removes two-thirds of the mass. The excised lobes do not grow back, but the remaining two small lobes grow rapidly to attain the original mass of the liver within 14 days after operation. When discrete lobes of the liver are removed by PH, there is no cell damage, and consequently cell proliferation occurs in the absence of cell death, fibrosis, or inflammatory events (Webber and Fausto 1994). Other kinds of damage do involve an inflammatory phase, but the regenerative outcome is the same.

1.3.1.1
Proliferation Capacity and Kinetics

The Greek legend of Prometheus describes a prodigious capacity of the liver for regeneration (Goss 1991). It is doubtful that the ancient Greeks knew that the liver really does regenerate, but the legend has been validated, over 2,000 years later, by experimental studies. The rat liver regenerates each time after 12 sequential partial hepatectomies (Michalopoulos and DeFrances 1997). The hepatocyte has enormous clonogenic potential. Mice with hereditary tyrosinemia type I, a fatal recessive liver disease caused by a deficiency of fumarylacetoacetate hydrolase (FAH), which leads to the accumulation of a hepatotoxic metabolite of tyrosine, were rescued by injecting as few as 1,000 normal hepatocytes into their livers. Furthermore, serial transplantations of hepatocytes could rescue tyrosinemic mice through four generations.

Calculations suggest that a single hepatocyte can divide at least 34 times, giving rise to 1.7×10^{10} cells, enough clonogenic capacity to regenerate about 56 rat livers of 3×10^8 hepatocytes each. This capacity for proliferation is remarkable, given the fact that most mature hepatocytes have a 4C ploidy, with some having even higher ploidy numbers, and the expansion is accomplished while maintaining all differentiated functions (Michalopoulos and DeFrances 1997).

All the cell types of the liver proliferate after PH, though the kinetics of DNA synthesis for each cell type are different. Hepatocyte DNA synthesis is initiated 10–12 h after partial hepatectomy and peaks at 24 h after PH, whereas DNA synthesis in biliary ductular cells, Kupffer cells, and Ito cells begins later and peaks later. The endothelial cells of the sinusoids begin proliferating last, reaching a peak of DNA synthesis at 4 days. In a young adult rat, up to 95% of hepatocytes divide at least once in restoring the original liver size. In older animals, liver regeneration is slower, less complete, and involves the proliferation of fewer hepatocytes (Michaelopoulos and De Frances 1997).

1.3.1.2
Gene Activity of Proliferation

Partial hepatectomy triggers a series of regulatory events that superimpose a pattern of transcriptional activity characteristic of proliferation onto the hepatocyte-specific transcriptional pattern (Webber and Fausto 1994; Michaelopoulos and De Frances 1997; Trembly and Steer 1998). Beginning at 30 min and continuing over 3 h, the activity of over 70 "immediate-early" genes is induced (Taub 1996). Induction of this group of immediate-early genes is independent of new protein synthesis; i.e., requires only the activation of existing transcription factors for their induction. The most important of these transcription factors are STAT3 (signal transducer and activator of transcription-3), which belongs to the JAK–STAT pathway of signal transduction, and partial hepatectomy factor (PHF/NF-κB), a liver-specific form of nuclear factor (NF)-κB. Many of the immediate-early genes contain binding sequences for STAT3 and NF-κB in their promoters and themselves code for transcription factors involved in initiating the early G_1 phase of the cell cycle, such as the protooncogene c-Myc and the leucine zipper proteins c-Jun, JunB, c-Fos, and liver regeneration factor (LRF)-1, which form DNA-binding complexes of the activator protein (AP)-1 type. High levels of these complexes are detected for several hours after the G_0 to G_1 transition (Taub 1996).

A second set of genes, called "delayed-early" genes, is induced independently of new protein synthesis starting at about 4 h after PH (Webber and Fausto 1994; Trembly and Steer 1998). Delayed-early genes code for proteins that take the cells through the remainder of the cell cycle. Important proteins synthesized are (1) the cyclins and cyclin-dependent kinases (cdks), which phosphorylate proteins such as the retinoblastoma protein (Rb) involved in passage from G_1 into S; (2) p53, a transcription factor that activates *p21*, which encodes a protein that inhibits cyclin/cdk activity, and (3) Bcl-X and Bcl-2, which, along with Rb, protect cells against apoptosis while going through the cell cycle.

Rb exhibits peaks in expression at 12, 30, and 72 h, with expression at 30 h. representing a greater than 100-fold increase over expression in non-regenerating liver. DNA synthesis in the livers of transgenic mice overexpressing *p21* is less than 15% of normal after PH and the mice have correspondingly decreased liver mass. Expression of *Bcl-2* and *Bcl-X* occurs at 6 h following PH. *Bcl-2* transcripts are expressed by non-hepatocyte cells at a 2-fold level above normal, whereas *Bcl-X* transcripts are expressed by hepatocytes at a 20-fold level, but the proteins encoded by these genes do not fluctuate significantly during regeneration (Michaelopoulos and DeFrances 1997; Trembly and Steer 1998).

The pattern of liver-specific protein synthesis shifts slightly during regeneration. A number of proteins identical to those produced in the fetal liver appear in regenerating liver as the result of activation of immediate-early genes. These include α-fetoprotein, hexokinase, and fetal isozymes of aldolase and pyruvate kinase. In addition, the expression of several liver function proteins is upregulated to compensate for tissue loss, including albumin and several genes that encode proteins involved in glucose regulation and metabolism, such as glucose-6-phosphatase, insulin-like growth factor binding protein-1, and phosphoenolpyruvate carboxykinase (Taub 1996).

1.3.1.3
Initiation of Liver Regeneration

A major question is how the immediate-early gene response is initiated. Mitogenic signals appear in the blood after PH, as shown by the fact that liver tissue or hepatocytes transplanted to ectopic locations replicate DNA following partial hepatectomy of the host liver. Furthermore, hepatectomy of one member of a parabiosed pair of rats induces growth of the intact liver of the other member of the pair (Michalopoulos and De Frances 1997).

These mitogenic signals are now known to be growth factors (Webber and Fausto 1994; Fausto and Webber 1995; Fausto et al. 1995; Michalopoulos and DeFrances 1997; Trembly and Steer 1998). The tumor necrosis factor (TNF)/TNFR1 signaling pathway is important for the initiation of liver regeneration, as is hepatocyte growth factor (HGF), EGF and TGF-α. TNF-α and interleukin (IL)-6 play an essential role in the transition of liver cells from G_0 to G_1 ("priming", or the acquisition of competence). TNF-α exerts its effects by binding to its receptor, TNFR1. Knockout mice lacking either TNFR1 or IL-6 are deficient in their ability to regenerate liver. DNA synthesis is severely impaired in mice with a TNF-α receptor deficiency, and antibodies to TNF-α decrease DNA synthesis. In both cases, there is a failure to activate STAT3 and NF-κB and to increase the production of c-jun and AP-1. These deficiencies can be corrected by the injection of IL-6. IL-6 is secreted by Kupffer cells, and plasma concentrations reach high levels by 24 h after PH. In mice homozygous for deletion of the IL-6 gene, STAT3 activation, AP-1 activity, Myc, and cyclin D1 are all markedly reduced, and DNA synthesis is suppressed. Secretion of IL-6 is stimulated by TNF-α. Collectively, these observations suggest that IL-6 is a mitogen essential for priming that is regulated by TNF-α. The results suggest a signaling pathway in which partial hepatectomy induces expression of TNF-α followed by activation of NF-κB, which induces IL-6, causing activation of STAT3. Activation of STAT3 and NF-κB initiate immediate-early gene expression (Trembly and Steer 1998). However, liver regeneration proceeds to completion in IL-6-deficient mice, indicating that other pathways can compensate for the loss of IL-6 (Michalopoulos and De Frances 1997). Thus, although TNF-α and IL-6 are required for the *normal* pace of liver regeneration, they are not indispensable for initiating regeneration.

HGF and EGF/TGF-α are essential for progression of the liver cells through G_1 to S, and may also be able to initiate liver regeneration (Michalopoulos and De Frances 1997). The inactive precursor of HGF (pro-HGF) is found in many tissues, including the liver, where it is produced primarily by stellate and endothelial cells. Pro-HGF is activated by urokinase plasminogen activator (uPA). Activated pro-HGF forms a heterodimer that exerts its effects on cells through its receptor, c-met.

HGF is a potent mitogen for hepatocytes cultured in the absence of serum (i.e., in the absence of other growth factors) and is thus a "complete mitogen." The plasma concentration of active HGF rises over 20-fold within 1 h after PH (Mars et al. 1995). The mechanism of this rise may be related to release of pro-HGF from the ECM by matrix degradation. Injecting HGF into uninjured liver results in a weak proliferative response, but infusing collagenase into the liver prior to injection of HGF greatly magnifies the response, suggesting that degradation of the matrix around hepatocytes and other cells plays a role in initiating regeneration. Furthermore, within 1–5 min after partial hepatectomy, urokinase receptor activity rises due to translocation of the receptor to the hepatocyte plasma membrane (Mars et al. 1995). This results in increased uPA activity, which in addition to activating pro-HGF converts plasminogen to plasmin, activating MMPs. Large amounts of pro-HGF are bound to the liver ECM (Michaelopoulos and De Frances 1997). Thus, the release of pro-HGF by matrix degradation and its cleavage by uPA could explain the rapid rise in active plasma HGF after PH.

The angiogenic growth factor VEGF-A plays a role in elevating the production of HGF and IL-6 (LeCouter et al. 2003). VEGF-A stimulates hepatocyte division when injected in vivo, but stimulates hepatocyte mitosis in vitro only in the presence of sinusoidal epithelial cells. Injury activates two distinct pathways in liver endothelial cells that lead to elevated levels of HGF and IL-6 by upregulating hepatocyte VEGF-A production. VEGF-A binds to both the VEGFR-1 and 2 receptors on endothelial cells. Binding to VEGR-1 results in enhanced secretion of HGF and IL-6. Binding to VEGF-2 enhances endothelial cell proliferation, thus increasing the number of cells producing higher levels of the growth factors.

EGF is also a complete mitogen for hepatocyte proliferation in vitro. EGF mRNA increases by tenfold in the liver within 15 min after PH and is expressed in hepatocytes, endothelial cells, Kupffer cells, and Ito cells (Michalopoulos and De Frances 1997; Trembly and Steer 1998). Plasma EGF levels rise somewhat later than HGF levels after partial hepatectomy, but the rise is less than 30%. The number of EGF receptors doubles in the first 3 h. EGF protein accumulates inside hepatocytes, but is also synthesized by the salivary glands and released into the bloodstream, suggesting that it acts through both autocrine and endocrine mechanisms. The latter mechanism may be the more important. Removal of the salivary glands 2 weeks prior to partial hepatectomy reduced plasma EGF concentration by 50%, and abolished the increase in EGF level after hepatectomy (Lambotte et al. 1997). DNA synthesis and mitosis was also reduced by 50% when the salivary glands were removed at the time of, or within 3 h after, partial hepatectomy. Salivary gland removal 6 h or more after hepatectomy had no effect. Administration of exogenous EGF to sialoadenectomized rats restored normal regenerative activity. In general, diminished EGF levels delayed the regeneration response by 24 h, but the liver nevertheless was fully regenerated by 7 days. These observations suggest that EGF affects events immediately following HGF-promoted entry into G_1. Insulin and norepinephrine amplify the mitogenic signals of HGF and EGF by binding to the α1-adrenergic receptor (Michaelopoulos and De Frances 1997).

TGF-α also binds to the EGF receptor and is induced in hepatocytes within 2–3 h after PH, rising to a peak between 12 and 24 h (Michalopoulos and De Frances 1997; Trembly and Steer 1998). Enhanced expression of TGF-α in hepatocytes under the influence of the albumin promoter leads to sustained high levels of DNA synthesis and to tumor formation, implying that it plays a role in progression of hepatocytes through the G_1/S transition and beyond. However, there is only a small increase in plasma TGF-α protein after PH, despite a large increase in TGF-α mRNA. Furthermore, liver regeneration proceeds normally in mice carrying a homozygous deletion of the TGF-α gene. Thus, it is unclear whether TGF-α has an important role to play in liver regeneration insofar as it affects hepatocyte proliferation. However, along with other growth factors produced by hepatocytes of regenerating liver such as FGF-1 and VEGF-A, TGF-α may be important for inducing the rise in proliferation of other liver cell types.

The picture that has thus emerged is one in which a variety of signals are necessary for normal liver regeneration. TNF-α, IL-6, HGF, and EGF appear to play definitive roles in initiating regeneration, with other growth factors playing facultative roles. How events as rapid as the increase in uPA activity within 5 min and the activation of immediate-early genes within 30 min are triggered is not yet clear.

1.3.1.4
Stopping Proliferation

DNA synthesis in regenerating rat liver is complete by 72 h. Little is known, however, about the mechanism by which proliferation is shut down. TGF-β1, which is normally made by Ito cells, may play a role in preventing and cessation of proliferation (Michaelopoulos and De Frances 1997). In vitro, TGF-β1 inhibits the mitosis of hepatocytes. However, liver regeneration, though slowed, proceeds to completion in mice transgenic for increased TGF-β1 gene expression (Kopp et al. 1995), suggesting that other factors act synergistically with TGF-β1 to terminate hepatocyte proliferation.

1.3.1.5
Remodeling

When proliferation ceases, hepatocytes exist as clusters of 10–14 cells lacking sinusoids and ECM (Martinez-Hernandez and Amenta 1995). The normal cellular organization of the liver and its ECM is then re-established. Bax, a pro-apoptotic protein of the Bcl-2 family, is most abundant in regenerating liver at the time of apoptosis and reorganization and thus may be involved in regulation of cell numbers. Three liver cell types synthesize and secrete ECM molecules: hepatocytes, endothelial cells, and Ito cells. Ito cells invade the hepatocyte clusters and synthesize laminin. The laminin is not organized into a basement membrane, perhaps because of a lack of entactin. It may serve as a stimulus for the subsequent invasion of endothelial cells that separate the hepatocytes into trabeculae two cells wide. In this manner, the sinusoidal spaces and space of Disse are re-established. Simultaneously, the other ECM molecules characteristic of normal liver are synthesized and deposited (Martinez-Hernandez and Amenta 1995).

1.3.1.6
Liver Regeneration via Stem Cells

The liver cannot regenerate by compensatory hyperplasia after toxic chemical injury that destroys the ability of hepatocytes to proliferate. Liver regeneration under these circumstances is accomplished by small oval cells that arise in the epithelium of the terminal bile ductules, proliferate, and differentiate into hepatocytes (Sell 2001; Webber and Fausto 1994; Alison et al. 1996; Thorgeirsson 1996; Grompe and Feingold 2001). The cells in the bile ductule epithelium are considered stem cells and the oval cells to represent a transit amplifying population. Oval cells are bipotential, and can differentiate into hepatocytes or bile duct epithelium (Sirica 1995). They express many of the markers characteristic of embryonic hepatoblasts, such as α-feto-protein, γ-glutamyl transpeptidase, cytokeratin 19, OC.2, OV-6, and Thy-1, but only a few of the markers expressed by hepatocytes, suggesting that they recapitulate the embryonic differentiation of hepatocytes (Thorgeirsson 1996; Hixson et al. 1997; Petersen et al. 1998). In addition, they express many of the markers expressed by bile ductule cells, in keeping with their probable origin from the terminal ducts (Fausto 1994; Thorgeirsson 1996).

Using antibodies against Thy-1.1, a 95%–97% pure population of oval cells has been isolated by FACS from rat liver treated with 2-acetylaminofluorene (2-AAF) and injured by CCl_4 or partial hepatectomy (Petersen et al. 1998). 2-AAF prevents hepatocyte proliferation, inducing a large oval cell response. These cells expressed the usual oval cell markers, but also expressed several hematopoietic cell markers (CD34, Thy-1, c-kit, flt-3 receptor), suggesting a potential overlap in the oval cell and hematopoietic stem cell phenotypes (Fujio et al. 1994; Petersen et al. 1998).

1.3.2
Pancreas

Type I diabetes is caused by an autoimmune response to glutamic acid decarboxylase (GAD), an enzyme produced by the β-cells of the islets (Yoon et al. 1999), leading to β-cell destruction, fibrosis of the pancreas, and near-total insulin deficiency. Type 2 diabetes is caused by resistance to insulin action, followed by deficiency of insulin secretion (Taylor 1999). Together, these two types of diabetes afflict 5% of the population, with type 2 accounting for over 90% of the cases. The damage to cardiovascular, renal, and neural systems caused by diabetes is severe, and wound healing is impaired.

The mammalian pancreas has some regenerative capacity (Slack 1995; Kritzik and Sarvetnick 2001). Acinar tissue regenerates rapidly after its selective destruction by temporary ligation of the pancreatic duct or ethionine treatment. These treatments spare the islet cells and pancreatic ducts. Prolonged ligation results in the permanent destruction of both acinar and islet cells, and irreversible fibrosis. Treatment with streptozotocin selectively ablates the β-cell population, but there is no significant recovery. Surgical removal of 90%–100% of the pancreas leads to some regeneration of acinar and islet tissue by two routes. The first is compensatory hyperplasia of acinar and islet cells, but this is very limited (Slack 1995; Kritzik and Sarvetnik 2001). The second is by regeneration from remnants of the pancreatic ducts, an observation that dates back to the 1920s (Slack 1995).

There is substantial evidence that regeneration from the ducts is accomplished by the proliferation of stem cells (Bonner-Weir et al. 1993). In mice transgenic for the interferon (IFN)-γ gene linked to an insulin promoter, the IFN-γ protein induces a severe inflammatory response in the pancreas that destroys islet cells and stimulates their regeneration from duct cells. BrdU labeling studies showed that the duct cells have a high proliferative activity. Immunohistochemical staining for insulin, carbonic anhydrase II (a duct cell antigen), and amylase (an exocrine enzyme) showed that these cells were insulin-positive and co-expressed carbonic anhydrase II and amylase, indicating their derivation from a ductal cell population (Gu and Sarvetnick 1993; Gu et al. 1994). In addition, a number of transcription factors known to be necessary for normal pancreatic development, such as PDX-1, HNF3β, Isl-1, and Pax-6, are expressed in the duct cells of the INF-γ transgenic mouse (Kritzik et al. 2001). Two populations of PDX-expressing cells are seen, one that expresses PDX-1 alone, and another that expresses PDX-1 and insulin. These are believed to represent earlier and later precursors of new islet cells, respectively.

Further evidence for the ability of the pancreas to regenerate by ductal stem cells has been obtained from mice made diabetic by streptozotocin treatment, followed by pancreatectomy (Hardikar et al. 1999). Mice treated with streptozotocin alone did not regenerate β-cells, whereas streptozotocin-treated mice that underwent 50% pancreatectomy became normoglycemic. Histological sections of the pancreas in such animals showed the regeneration of islets from the pancreatic duct. Finally, the reversal of diabetes in non-obese diabetic (NOD) mice by the induction of TNF-α expression combined with re-education of newly differentiating T cells with self-antigens was able to restore normoglycemia in up to 75% of the animals after treatment was discontinued (Ryu et al. 2001). While the mechanism of recovery in these mice is not known, the interruption of autoimmunity might rescue surviving β-cells and/or stimulate the regeneration of new β-cells from ductal precursors.

1.4
Regeneration of Mesodermal Derivatives

Several mesodermal tissues exhibit physiological regeneration or regenerate after injury via stem cell proliferation. Blood and immune cells are continually replenished from hematopoietic stem cells (HSCs) in the bone marrow. Fractured bone regenerates from mesenchymal stem cells (MSCs) residing in the periosteum of the bone and in the marrow cavity. Skeletal muscle regenerates after small tears by budding and after larger injuries via satellite cells (SCs) associated with the individual myofibers.

1.4.1
Skeletal Muscle

Skeletal muscles are composed of multinucleate myofibers organized into fascicles that are grouped together to form individual muscles. Myofibers are organized into fascicles surrounded by perimysia, and the fascicles collectively constitute the muscle, which is surrounded by an epimysium. Each myofiber is surrounded by a basement membrane made by its connective tissue sheath, the endomysium. Other muscle ECM components synthesized by the endomysium are sulfated proteoglycans, including a muscle-specific sulfated proteoglycan (Caplan 1991), and tetranectin, a matricellular protein that is not a structural component of the ECM, but serves as an interactive agent with other ECM proteins, cell surface receptors, cytokines, and proteases (Wewer et al. 1998). Tetranectin binds sulfated polysaccharides, suggesting it might interact with glycosaminoglycans (GAG) chains of PGs. It is particularly prominent at myotendinous junctions, considered to be the equivalent of focal adhesion sites in muscle (Wewer et al. 1998).

1.4.1.1
Origin of Regenerated Myofibers

Vertebrate skeletal muscle in the neonates and adults of virtually every species examined contains a population of stem cells called "satellite cells" located between the sarcolemma and the overlying basement membrane. Satellite cells were first identified by electron microscopy in frog skeletal muscle (Mauro 1961). DNA labeling studies subsequently showed them to be the source of regenerated muscle after injury in several species (Hinterberger and Cameron 1990). Satellite cells represent between 1% and 5% of the total nuclei in adult mammalian muscles (Allbrook 1981). Satellite cells separated from chicken pectoral muscle by Percoll density centrifugation and cultured in vitro give rise to large clones of cells that form myofibers (Yablonka-Reuveni et al. 1987). Experiments implanting SCs into dystrophic muscles showed that not only do the implanted cells contribute to the regeneration of myofibers, they also give rise to more SCs, indicating that they are self-renewing (Blaveri et al. 1999).

Several markers expressed at high levels specifically define quiescent SCs: (1) myocyte nuclear factor (Garry et al. 1997), which may prevent transcription of the muscle regulatory factors (MRFs) that promote differentiation into myofibers; (2) the c-met tyrosine kinase receptor (Cornelison and Wold 1997), which plays a key role in the activation of SCs (see below); (3) p130, a protein that blocks cell cycle progression by binding to E2F transcription factors and also inhibits the differentiation of cultured mouse C2 myoblasts by inhibiting the expression of MyoD (Carnac et al.

2000); and (4) syndecan 3 and 4, transmembrane heparan sulfate proteoglycans (HSPGs) (Cornelison et al. 2001).

Satellite cells are themselves the product of a "side-population" of multipotent muscle stem cells (MuSCs) that has both hematopoietic and myogenic developmental potential and originates in the bone marrow (McKinney-Freeman et al. 2002). The existence of MuSCs was revealed by genetic studies on mice null for the Pax7 gene. These mice exhibit normal muscle development, but lack satellite cells. Nevertheless, the muscles of Pax7-null mice were shown to harbor a population of CD45+ stem cells that cannot differentiate into muscle, but can differentiate into hematopoietic lineages when engrafted into bone marrow (Seale et al. 2000). CD45+ cells isolated from uninjured muscle of wild-type mice behave the same way; they are unable to form muscle but can rescue the hematopoietic system when injected into irradiated mice (Jackson et al. 1999). However, CD45+ cells from injured wild-type muscle are able to regenerate muscle (Seale and Rudnicki 2000). These observations suggest that production of satellite cells during embryogenesis requires the expression of Pax7 and that under injury conditions in adults, MuSCs replenish the supply of satellite cells used for muscle regeneration. Recently, it was shown that injured muscle produces several isoforms of Wnt signaling proteins and that Wnt signaling induces the expression of Pax7 and the myogenic specification of CD45+ MuSCs (Polesskaya et al. 2003). MuSC proliferation and myogenic specification was significantly reduced by injection of the Wnt antagonists sFRP2 and 3 into regenerating muscle.

1.4.1.2
Cellular and Molecular Events of Muscle Regeneration

Satellite cell proliferation is triggered by muscle injury, exercise, or weight-bearing (Grounds 1991), and has also been noted in astronauts after space flight (Pastoret and Partridge 1998). Muscle does not regenerate across a gap and excised segments of muscle are not regenerated. Incisions through myofibers that do not remove tissue and leave the cut ends of the myofibers apposed, evoke regeneration by cytoplasmic budding and rejoining of the cut fibers. Such local injuries can be created experimentally by small transverse cuts or laser. Most muscle injuries, however, are more extensive and result in tissue death. There are several experimental animal models of such injury, each of which has its advantages (Carlson 2003). Genetic myopathies such as muscular dystrophy, or administration of myotoxins damages only myofibers, while leaving vascular and neural fibers intact. Relatively localized ischemic injuries can be created by crushing, cutting, freezing, and vascular clamp. Extensive ischemic degeneration can be achieved by removing the muscle and grafting it back to the muscle bed (free graft), or mincing the muscle and grafting the mince back to the muscle bed (Carlson 1983, 2003; Pastoret and Partridge 1998). The extensor digitorum longus or gastrocnemius muscles are most often used in studies of rat muscle regeneration.

Regeneration of the free-grafted extensor digitorum longus muscle has been well described in both amphibians and mammals (Hansen-Smith and Carlson 1979; Carlson 2003). The first event to occur after free-grafting a rat muscle is a wave of necrosis that sweeps from the periphery of the graft to the center over a 7-day period.

Myofibers are destroyed by complement activation and by calcium influx that activates calcium-dependent neutral proteases. As the wave of necrosis sweeps through the muscle, a typical inflammatory response follows, in which soluble chemoattractants attract neutrophils and macrophages into the degenerating muscle (Grounds and Yablonka-Reuvini 1993; Pastoret and Partridge 1998). The macrophages promote proliferation of satellite cells by the production of PDGF, TGF-β, FGF-2, and leukemia inhibitory factor (LIF) and stimulate neovascularization via secretion of angiogenic factors (Grounds and Davies 1996; Pastoret and Partridge 1998). Zymographic studies indicate that MMP-9 (92-kDa type IV collagenase; gelatinase B) is upregulated during the inflammatory phase; in situ hybridization showed that the mRNA for this enzyme is localized to satellite cells (Kherif et al. 1999). The breakdown of ECM components by MMP-9 may serve not only to digest products of necrosis, but also to release growth factors important for SC proliferation that are bound to ECM molecules.

The basement membrane of the degenerating myofibers is fragmented due to loss of fibronectin, laminin, heparan sulfate proteoglycan, and collagen IV, allowing the detachment of satellite cells from the lamina (Gulati et al. 1983). MMP-2 (72-kDa type IV collagenase; gelatinase A) may be involved in breaking down the basement membrane. Cultured C2C12 myoblasts and normal skeletal muscle synthesize MMP-2, the expression of which is upregulated during the period of maximal SC proliferation and fusion into myotubes (Kherif et al. 1999). As the new myofibers differentiate, a continuous basement membrane is synthesized around them by fibroblasts (Hansen-Smith and Carlson 1979). Contractile proteins are expressed and the regenerated myofibers exhibit slow spontaneous contractions toward the end of the first post-transplantation week. Reinnervation of the regenerating muscle begins during the second week post-transplantation, and the speed of contraction subsequently increases until it approaches normal at 30–40 days after transplantation (Carlson and Gutmann 1972).

The process of satellite cell proliferation, fusion, and differentiation during regeneration appears to largely recapitulate myogenesis during embryogenesis. There is no difference in the in vitro growth rates and fusion characteristics between embryonic myoblasts and satellite cells, and both exhibit the same increase in creatine kinase activity and shift in creatine kinase isozyme profile as myofiber formation proceeds (Jones 1982). Like myoblasts of embryonic muscle, satellite cells of regenerating mouse muscle express M-cadherin (Moore and Walsh 1993). Normal adult chicken muscle synthesizes only the small heparan sulfate and dermatan sulfate proteoglycans, whereas regenerating muscle reverts to the embryonic pattern of abundant synthesis of large chondroitin sulfate proteoglycans (Carrino et al. 1988). Embryonic and regenerating myofibers of chicken muscle both synthesize the embryonic myosin heavy chain, the α fast light myosin chain, and β-tropomyosin (Gorza et al. 1983; Matsuda et al. 1983; Dix and Eisenberg 1991). This is true of myofibers derived from SCs of either fast (pectoralis major) or slow (anterior latissimus dorsi) muscle. Muscle cells also contain the S1 protein, a variant of EF-1α, which is ubiquitous to all cells (Khalyfa et al. 1999). The ratio of EF-1α to S1 is very high in embryonic muscle, but is low in adult muscle. Regenerating muscle reverts to the embryonic ratio. However, increased expression of EF-1α is also observed in cells undergoing apoptosis (Duttaroy et al. 1998); thus, the increase in EF-1α in regenerating muscle may reflect myofiber death as well as the events of regeneration. Tetranectin is expressed by the satellite cells, myotubes, and the stumps of damaged myofibers (Wewer et al. 1998).

Quiescent satellite cells from old (9-month) rats do not express any of the MRFs active in the embryonic development of muscle (Koishi et al. 1995; Megeny et al. 1996), but express all of them in culture (Smith et al. 1994). The sequence of expression, however, is different from the sequence seen in embryonic development. MyoD is expressed first, at 12 h after plating. Myf5 and MRF 4 expression sets in at 48 h and is correlated with the first division cycle. Myogenin expression is detected sporadically at 48 h, and in all cultures at 72 h, coincident with the ability to immunostain cultures for sarcomeric myosin heavy chain and 24 h before myoblast fusion, consistent with the role ascribed to myogenin in myogenic differentiation (Smith et al. 1994). A slightly different expression pattern of MRFs was observed when their mRNAs were assayed in single activated satellite cells of cultured individual myofibers via single-cell polymerase chain reaction (Cornelison and Wold 1997). Either MyoD or myf5 were expressed first. Most cells then expressed myf5 and MyoD simultaneously, followed by myogenin expression in cells expressing both myf5 and MyoD. Ultimately, many of the cells expressed myf5, MyoD, myogenin, and MRF4 simultaneously. Thus, there may be a great deal of heterogeneity of expression of MRFs among activated satellite cells that may be masked by the "average" pattern.

1.4.1.3
Regulation of Muscle Regeneration by Growth Factors

Satellite cell proliferation is promoted by macrophage-produced cytokines during the inflammation phase of muscle regeneration. These cytokines, however, do not appear to be absolutely essential for activation of SCs. Abolishing the inflammatory response in rats by total body irradiation reduces the efficiency of regeneration in muscles injured by incision or crush, but does not abolish regeneration (Robertson et al. 1992). By contrast, muscles subjected to heavy local irradiation, which prevents the proliferation of SCs but not the inflammatory response, fail to regenerate.

Several lines of evidence suggest that, like hepatocytes, SCs are activated following injury by HGF. First, SCs of cultured myofibers can be selectively induced to enter the cell cycle earlier than normal by exposing them to a fraction of crushed muscle extract (Bischoff 1986). Second, HGF is the only growth factor that can mimic the active fraction of extract (Allen et al. 1995). Third, the ability of crushed muscle extract to stimulate the early entry of SCs into the cell cycle is abolished by anti-HGF antibodies (Tatsumi et al. 1998). Fourth, although HGF is not expressed in uninjured adult myofibers, it is present in the ECM surrounding the muscle fibers, is expressed by proliferating SCs, and is downregulated upon their fusion and differentiation into myotubes (Jennische et al. 1993; Tatsumi et al. 1998). Fifth, the receptor for HGF, c-met, is present in the plasma membrane of quiescent SCs (Allen et al. 1995; Cornelison and Wold 1997; Tatsumi et al. 1998) and is co-localized with HGF in the activated SCs of regenerating muscles in *mdx* dystrophic mice (Tatsumi et al. 1998). Lastly, SC activation is stimulated by the injection of HGF into the uninjured tibialis anterior muscle of 12-month-old rats (Tatsumi et al. 1998). These observations suggest that injury releases HGF from the ECM of the muscle cells, triggering SC activation and proliferation. As part of the activation process, the SCs themselves express HGF, which acts in an autocrine fashion. It is likely that the HGF bound to muscle

ECM is activated in the same way as it is in regenerating liver, by the uPA→plasminogen activator→plasmin cascade (Miyazawa et al. 1996). Presumably, some of the HGF produced by proliferating SCs would become bound to newly synthesized ECM molecules of the regenerating muscle, to be available for release in any subsequent round of regeneration.

Several other growth factors found in crushed muscle extract can modulate the proliferative activity of SCs in vitro (Johnson and Allen 1995; Grounds and Davies 1996; Pastoret and Partridge 1998). In uninjured muscle, they are bound to ECM components and their receptors are not expressed by quiescent SCs (DiMario et al. 1989). Following injury, these growth factors are released to interact with upregulated receptors on activated SCs. The regenerating muscles of *mdx* dystrophic mice exhibit abnormally high levels of FGF in the endomysium (Anderson et al. 1991), perhaps because of high levels of MMPs that are degrading the ECM (Kherif et al. 1999). The muscles of these mice also express LIF and IL-6. LIF binds to ECM components and stimulates the formation of larger myotubes when infused continuously into *mdx* muscle, presumably by increasing the proliferation of activated SCs (Kurek et al. 1996). PDGF-BB, FGF-2, and transferrin are also present in crushed muscle extract (Chen et al. 1994) but neither they nor IGF-1, IGF-2, TGF-β1, or TGF-β2 can activate quiescent satellite cells (Johnson and Allen 1995). High-affinity FGF-2 receptors are not present in quiescent satellite cells, but are upregulated by HGF in vitro and in vivo following SC activation, coincident with the expression of FGF-2 (Garrett and Anderson 1995). It appears that HGF and FGF-2 together promote SC proliferation, whereas IGF I and II and TGF-β2 suppress proliferation and promote myotube formation (McLennan and Koishi 1997).

Heparan sulfate is required for the activation of SCs by HGF or FGF, perhaps by promoting dimerization of their receptors (Rapraeger 2000). Inhibition of HSPG sulfation by treatment of intact myofibers with chlorate results in delayed proliferation and altered MyoD expression of SCs in injured muscle (Cornelison et al. 2001). This suggests that syndecan 3 and 4 expression by SCs may provide the heparan sulfate required for dimerization of HGF and FGF receptors during SC activation.

Normal muscle regeneration requires innervation, despite the fact that myofibers regenerate in the absence of nerves. Denervation prevents or retards the full structural and functional differentiation of regenerating muscle (Carlson and Gutmann 1975), including the specialized intrafusal fibers of the muscle spindles and sensory receptors that function in the stretch reflex and in the regulation of muscle tone (Rogers 1982). Whether the nerves supply a growth factor or factors to the muscle in addition to electrical activity is unknown.

1.4.2
Bone

1.4.2.1
Bone Regenerates via Mesenchymal Stem Cells

The regeneration of fractured bone is accomplished by MSCs in the periosteum, endosteum, and marrow stroma (McKibben 1978; Einhorn 1998). The existence of

MSCs in bone marrow was first indicated by the ability of marrow cells to form bone when transplanted to ectopic sites (Urist 1965; Urist and McLean 1962; Tavassoli and Crosby 1968). In vitro, MSCs are distinguished from HSCs by their adherence to the substrate. MSCs can differentiate clonally into bone, cartilage, and adipocytes in vitro or when implanted subcutaneously in diffusion chambers or ceramic blocks (Friedenstein et al. 1970; Owen 1987; Haynesworth et al. 1990; Lian et al. 1999; Pittinger et al. 1999). Their antigenic phenotype is negative for hematopoietic (CD14, CD34, and CD45) and endothelial surface antigens (P-selectin and von Willebrand factor VIII; Pittinger et al. 1999; Pittinger and Marshak 2001) and positive for a large number of growth factor, ECM, and other receptors. FACS-sorted bone marrow cells expressing the antigen STRO-1 have been isolated that differentiate into bone, cartilage, and adipocytes; this antigen may uniquely identify MSCs (Aubin and Liu 1996). MSCs in micromass pellet culture behave like mesenchymal cells of the limb bud in micromass culture, differentiating as cartilage in high-glucose medium (Mackay et al. 1998) and expressing the BMP-1B receptor (Chen et al. 1998).

Following the fracture of endochondral bones, blood vessels within and without the bone are torn, resulting in the formation of a clot (hematoma) in and around the break. Hypoxia results in osteocyte death for a limited distance on either side of the fracture. Platelets in the clot release PDGF and TGF-β, initiating an inflammatory phase in which the hematoma is invaded by neutrophils and macrophages (Einhorn 1998). Some macrophages become osteoclasts that degrade the matrix of the dead bone. The remaining phases of fracture repair appear to recapitulate the events of endochondral bone development. Within a few days, periosteal MSCs differentiate on both sides of the fracture to osteoblasts (hard callus) in a process of direct ossification (Brighton and Hunt 1991; Glowacki 1998). Within the fracture space itself, MSCs in the periosteum and bone marrow proliferate to form a "soft callus." These MSCs condense and differentiate into chondrocytes that secrete cartilage-specific matrix. The chondrocytes then undergo hypertrophy and die concurrent with matrix calcification. The calcified matrix is then invaded by periosteal blood vessels, induced by angiogenic factors produced by the hypertrophic chondrocytes (Glowacki 1998). The invading blood vessels are accompanied by MSCs that differentiate into osteoblasts, which replace the cartilage matrix with bone matrix. Synthesis of matrix components such as type II, I, IX, and X collagens, aggrecan, and fibronectin of cartilage and osteonectin, osteopontin, and osteocalcin of bone appear to follow patterns identical to those in developing and growing bones (Einhorn 1998).

1.4.2.2
Regulation of Bone Regeneration by Growth Factors

The differentiation of cartilage and bone during fracture repair is controlled by the same growth and transcription factors involved in endochondral bone development. Some of the growth factors are synthesized by the chondrocytes and osteoblasts of the regenerating bone, while the degradative activity of osteoclasts releases others, such as TGF-β, from the organic components of the dead bone matrix (Pfeilschifter et al. 1986, 1987). As the chondrocyte callus matures and is replaced by bone, transcripts of genes that encode proteins active in chondrocyte and osteoblast differentiation, such as bone morphogenetic proteins (BMPs), Indian hedgehog (Ihh), Cbfa1 (Runx) and osteocalcin, are detected (Ferguson et al. 1998).

Members of the TGF-β family of growth factors are particularly important for chondrogenesis and osteogenesis in fractured bone. BMPs are released from degrading bone matrix after fracture, and they and their receptors are strongly expressed by soft callus cells (Bostrom 1998). BMPs were detectable by antibody staining in cells of the soft callus as early as 3 days after fracture in a rabbit mandible model. Antibody staining increased as cartilage differentiated, and by 2 weeks after fracture osteoblasts were stained (Jin and Yang 1990; Yang and Jin 1990). BMP-4 transcripts were expressed in mesenchymal cells of the periosteum and marrow during the early phase of rib fracture repair, then disappeared when chondrogenesis began (Nakase et al. 1994). Monoclonal antibodies to BMP-2, 4, and 7 showed an increasing intensity of staining for these BMPs in periosteal mesenchymal cells in the region of hard callus formation and in the proliferating mesenchymal cells of the early soft callus, and chondroblasts differentiated from these cells. Less intense staining was seen in maturing and hypertrophic chondrocytes, but staining was again intense in osteoblasts replacing the cartilage with bone (Bostrom 1998). A mutation in the mouse *short ear* gene, which encodes BMP-5, results in congenital bone defects and a reduced capacity to repair fractures, suggesting that this BMP plays an important role in both bone embryogenesis and regeneration (Kingsley et al. 1992). BMPs are not expressed in uninjured bone, but immunolocalization studies show that BMPR-IA and IB are expressed in the periosteal cells of uninjured bone and are upregulated in these cells after fracture, parallel with the upregulation of BMPs (Bostrom 1998). Activin receptors are also expressed in proliferating and maturing chondrocytes in the fracture, but activin expression is weak, suggesting that activin receptors might act as receptors for BMPs (Bostrom 1998).

Northern blotting and immunolocalization studies indicate that high levels of TGF-β and FGF-1 are expressed during chondrogenesis of the soft callus, but not in the region of hard callus formation. TGF-β is present earlier in the hematoma and periosteum, but its source appears to be platelets and release from degrading bone matrix rather than synthesis by periosteal cells. Lower levels of PDGF and FGF-2 are expressed in the soft callus (Bolander 1992). Less is known about the expression and function of other molecules during fracture repair that are involved in the development of endochondral cartilage templates during ontogenesis. For example, it is not known whether the same upregulation of hyaluronidase and adhesion proteins (NCAM, fibronectin, Cyr61) seen in embryonic skeletal condensations (Knudson and Toole 1985; Oberlender and Tuan 1996; Gehris et al. 1997; Wong et al. 1997) is required for mesenchymal condensation within the soft callus, nor whether the Ihh-PTHrP and Delta/Notch signaling pathways (Lee et al. 1995; Vortkamp et al. 1996; Zou et al. 1997; Crowe et al. 1999) control the transition of proliferating chondrocytes to more mature and hypertrophying chondrocytes. However, it is likely that these molecules play similar roles in bone regeneration.

Young, growing animals heal fractures more rapidly and reproducibly than older animals. In rats, the decline in the ability to repair fractures with age is associated with a decline in the number of progenitor cells or their responsiveness to bioactive factors. A decrease with age in osteoprogenitor cells derived from bone marrow after fracture has been reported (Quarto et al. 1995), as has an age-related decline in the bone-inductive response of non-osteogenic tissues to BMP-2 implants (Fleet et al. 1997).

1.4.3
Blood and Lymphoid Cells

Blood consists of plasma, erythrocytes, and a myeloid component consisting of platelets, granulocytes (basophils, eosinophils, and neutrophils), macrophages, osteoclasts, and myeloid dendritic cells. Also circulating in the blood and lymph systems are cells of the immune system: B cells, T cells, natural killer cells, and lymphoid dendritic cells. Mature erythroid, myeloid, and lymphoid cells (with the exception of memory B and T cells), have half-lives of only days to weeks, and thus must be continually regenerated from multipotent HSCs residing in the stroma of the bone marrow. The regeneration of blood and lymphoid cells was one of the most intensely studied biological processes of the twentieth century, and continues to be so today (Weissman 2000).

The existence of long-term (LT) HSCs in bone marrow that generate blood and lymphoid cells while self-renewing has been demonstrated by bone marrow reconstitution experiments. In these experiments, mice and rats are lethally irradiated to eliminate the ability of bone marrow cells to divide, followed by injecting them with progressively diluted suspensions of labeled marrow cells (Harrison et al. 2001). If the injected marrow contains LT HSCs that give rise to all blood and immune cell lineages, this is revealed by the formation of colonies of erythroid and myeloid cells in the spleen and colonies of lymphoid cells in the thymus (Till and McCulloch 1980). Colony-forming assays have shown that a single transplanted cell (the HSC) is capable of homing to the bone marrow and giving rise to all blood and lymphoid cell types. Bone marrow cells diluted and serially transplanted from reconstituted animals to new irradiated hosts also give rise to complete blood and lymphoid lineages, indicating that the HSC renews itself. The existence of separate short-term (ST) blood and lymphoid stem cells downstream from the HSC (colony forming units, CFUs) has been demonstrated in the same way. HSCs divide to give rise to a common lymphoid progenitor (CLP) from which is derived all the cells of immune system, and a common erythroid/myeloid progenitor (CMP) which gives rise to the blood cell lineages (Broxmeyer and Williams 1988). While the "definitive" LT HSC has not been conclusively purified, the bone marrow cell fraction that is $CD34^+$ KDR^+ $c\text{-}Kit^+$ $Thy\text{-}1^{lo}$ Lin^- Sca^+ is heavily enriched in stem cells that can form all blood and lymphoid lineages (Spangrude et al. 1988; Ikuta et al. 1990; Ziegler et al. 1999; Melchers and Rolink 2001). These cells are small in size (6 μm), with dense chromatin, and are capable of differentiating in vitro into both lymphoid and myeloid lineages. They represent about one in every 10^5 bone marrow cells (Berardi et al. 1995).

Each step in the developmental pathways of the HSC and its progeny is dependent on combinations of growth factors produced by the stromal cells of the marrow. Some of these growth factors, such as stem cell factor (SCF) and IL-1 and 3, are common to several cell lineages, whereas others, such as erythropoietin, are specific for one or the other lineage. The intracellular signals generated by HSCs through interactions with the stromal environment lead to the up- or downregulation of specific sets of transcription factors that determine the patterns of gene activity defining each differentiation step. These transcription factors act through a variety of mechanisms that affect their binding to regulatory regions of genes (Orkin 2001). First, a single transcription factor may act as a dominant regulator. For example, the zinc-finger protein GATA-1 and the PU.1 protein induce blood cell lineages. The particular lineages induced may depend on the concentration of the transcription factor. Second, the activation of genes for one lineage is coupled to inactivation of the genes for another. For example, upregulation of cell surface antigens defining the blood cell lineages induced by GATA-1 and PU.1 is accompanied by downregulation of cell surface antigens characteristic of other lineages. Third, combinatorial positive and negative protein–DNA and protein–protein interactions among transcription factors and cofactors are probably the rule, even with dominant regulators. For example, GATA-1 must interact with a co-factor called "FOG" (derived from "friend of GATA-1") for normal erythropoiesis and platelet formation.

1.5
Developmental Potential of Adult Stem Cells

The cell types normally regenerated by adult stem cells under physiological or injury conditions are called their prospective significance or fate. The prospective fate of a stem cell is normally restricted by local signals to a subset of phenotypes within the germ layer of origin. However, there is evidence that when exposed to signals they

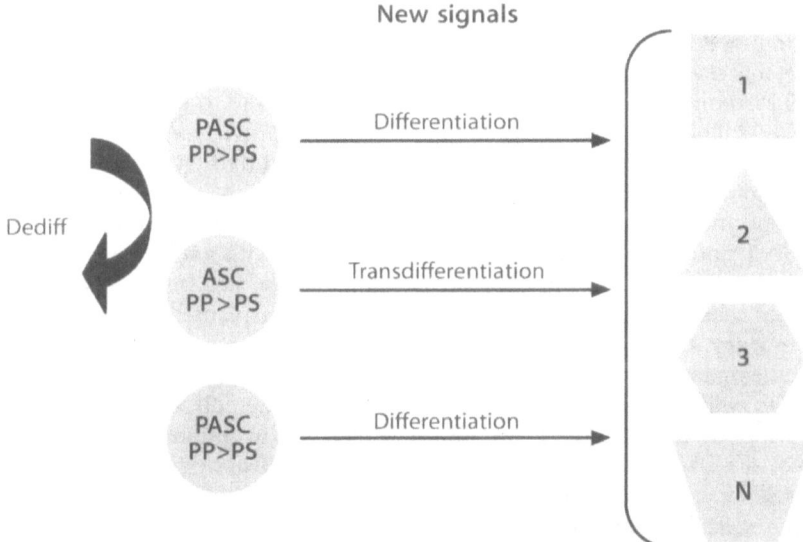

Fig. 2 Possible explanations for the wide range of differentiation exhibited by stem cells when tested in chimeric embryo, bone marrow reconstitution, and in vitro assays, where they are exposed to new sets of signals. 1, 2, 3, N represent differentiated cell types. Middle, an adult stem cell (ASC) whose prospective potency is greater than its prospective significance undergoes direct transdifferentiation without reverting to a less differentiated state. The currently active transcription program is shut off as the new transcription program is activated by the new signaling environment. Top, an adult pluripotent stem cell (PASC) that has gained pluripotency by dedifferentiation and can now differentiate into any cell type. Bottom, pluripotent adult stem cells (PASC) exist as such in tissues and respond by differentiating into any cell type. PP >PS indicates that prospective potency is greater than prospective significance

normally would not see, adult reserve stem cells have the ability to give rise to cell types other than the ones they would normally produce. That is, they have a prospective potency greater than their prospective fate. Differentiation of a cell into cell types other than its prospective significance or origin is commonly called transdifferentiation (Fig. 2).

To test the prospective potency of ASCs, they are first labeled in some way and then challenged in an assay that exposes them to foreign signals. Cells can be labeled with transgenes for markers such as β-galactosidase or GFP, by incorporation of lipophilic dyes, or BrdU incorporation. Natural markers such as the Y chromosome, or species-specific DNA sequences and antigens are also used. Assays for transdifferentiation include injecting ASCs into irradiated or *scid* host mice (bone marrow reconstitution assay) or into early embryos (chimeric embryo assay), or culturing ASCs with other cell types in vitro. The presence of cells containing the donor label(s), but having the morphology, molecular identity, and function of differentiated cells from other lineages is considered evidence that the cells are multipotent or pluripotent.

1.5.1
Neural Stem Cells

Clonally derived NSCs from ROSA26 mice transgenic for the *lacZ* gene were tested in the bone marrow reconstitution assay using Balb/c mice as hosts (Bjornson et al. 1999). The results strongly suggested that NSCs were able to transdifferentiate into myeloid, lymphoid, and early hematopoietic cells. The *lacZ* gene was detected in the splenic DNA of injected animals. Cells could be isolated by flow cytometry from the bone marrow, spleen, and peripheral blood of host animals, using antibodies to the ROSA26-specific cell surface antigen H-2k^b. Flow cytometry, using antibodies to CD3e, CD19, and CD11b, respectively, in combination with antibodies to H-2K^b, revealed the presence of T lymphocytes, B lymphocytes, and myeloid cells derived from donor cells. Clones of bone marrow cells from transplanted animals exhibited a high frequency of β-galactosidase activity. These colonies included granulocyte, granulocyte–macrophage, macrophage, megakaryocyte, and B cells. Pluripotency of clonally derived neurospheres from ROSA26 NSCs was demonstrated in the chimeric embryo assay by injecting them into mouse blastocysts or into the prospective amniotic cavity of early chick embryos (Clarke et al. 2000). Twenty-five percent of recipient chick embryos and 12% of recipient mouse embryos showed integration of β-gal expressing cells into the tissues and organs of all germ layers. Non-neural tissues formed by donor cells were epidermis, notochord, mesonephros, somites, heart muscle, lung epithelium, stomach and intestinal epithelium, stomach and intestinal wall, and liver. Donor cells expressed the molecular markers typical for these tissues. The frequency of chimerism ranged from 38% (heart) to 96% (intestinal epithelium).

Several experiments in vitro have demonstrated the differentiation of NSCs into muscle. Purified (FAC sorted) mouse NSCs (PNA^{lo} HSA^{lo}) transfected with the GFP gene and co-cultured with C2C12 myoblasts differentiated in vitro into skeletal myotubes expressing the muscle markers α-actinin-2 and myosin heavy chain (Rietze et al. 2001). Fifty-seven percent of the plated NSCs exhibited muscle characteristics.

GFP-labeled NSCs differentiated into cardiac muscle cells when co-cultured with cardiomyocytes from neonatal rats (Condorelli et al. 2001). Clarke et al. (2000) cultured ROSA26 NSCs with mouse embryoid bodies to evaluate the capacity of inductive signals from ESCs to guide the differentiation of the NSCs. The ROSA26 cells harbored a neomycin resistance gene that allowed the elimination of the G418-sensitive ES cells from the cultures after inductive signaling had occurred. Progeny of the NSCs expressed desmin and many cells fused to form muscle-like syncytia that were immunoreactive to myosin heavy chain.

The conversion of NSCs into other cell types has also been demonstrated in the regenerating axolotl tail. Since the tail of recently hatched axolotl larvae is thin and transparent, the fate of labeled NSCs can be followed live with a light microscope. Echeverri and Tanaka (2002) inserted the GFP gene into individual ependymal cells of the axolotl spinal cord after tail amputation. The gene was under the control of the promoter for GFAP, which is expressed only in the ependymal cells. Staining with GFAP antibody showed that all the GFP-expressing cells were ependymal cells. The labeled cells divided and, as expected, their progeny differentiated into neurons and glia in the regenerating spinal cord, as well as neural crest derivatives. Some NSCs, however, migrated out of the ependymal tube and transdifferentiated into muscle, which was identified with antibody to muscle-specific myosin heavy chain, and into chondrocytes, at frequencies of 24% and 12%, respectively. An unanswered question is whether blastema cells derived from cartilage, muscle, and connective tissue transdifferentiate into neurons.

1.5.2
Hepatic Oval Cells

Oval stem cells of rat liver bile ductules have been shown to be bipotential, capable of giving rise to hepatocytes and bile duct cells (Petersen et al. 1998). These cells were isolated and purified by cell sorting for the Thy-1.1^+ population (purity >95%), cultured for 6 months in serum-free medium, and then tested for their ability to transdifferentiate into insulin-producing pancreatic cells by growing them in medium supplemented with 10% fetal calf serum and a high (23-mM) glucose concentration (Yang et al. 2002). After about 2 months of culture, the cells formed small, spheroidal, islet-like clusters that enlarged and became connected by ductal structures. These clusters expressed transcripts for a number of endocrine cell differentiation markers and hormones, such as the transcription factors PDX-1, Nkx2.2, Nkx6.1, PAX-4 and PAX 6, the cell surface marker Glut-2, and insulin I, insulin II, glucagons, and somatostatin. Immunocytochemistry showed that the cells were producing glucagon and insulin. When stimulated with glucose, the cells synthesized and secreted insulin, a response that was enhanced by nicotinamide. A preliminary study indicated that the hyperglycemia of streptozotocin-induced diabetic NOD/*scid* mice was normalized by implants of the transdifferentiated cells under the renal capsule.

1.5.3
Satellite Cells

Regeneration of infarcted cardiac muscle has been reported after transplanting *lacZ* and 4'-6'-diamidino-2-phenylindole (DAPI_-labeled autologous satellite cells obtained from the canine tibialis anterior muscle into the infarct region (Kao et al. 2000). Antibody staining for connexin-43 revealed the presence of gap junctions in β-galactose[+] cells. The labeled cells were reported to take on the morphological characteristics of cardiomyocytes. Scar formation was reduced and systolic function of the engrafted hearts was improved compared to controls. It was not entirely clear, however, that transdifferentiation of SCs into cardiac muscle cells took place in this experiment. False-positive staining for LacZ product occurs when inflammation is present, as it would be in this case, and DAPI is not an ideal labeling reagent.

Satellite cells also appeared able to reconstitute the hematopoietic system in a bone marrow repopulation assay (Gussoni et al. 1999). These cells were a FACS-purified, c-kit[−] CD45[−] population and thus were not the muscle stem cells that have hematopoietic potential. Of male donor cells, 7,000–20,000 were injected via tail vein into lethally irradiated *mdx* female mice. Thirty days later, the muscle and bone marrow of the recipients was examined for the presence of the Y chromosome. Three to nine percent of the recipient tibialis anterior muscle contained Y[+] nuclei and 30%–90% of the bone marrow cells contained Y[+] nuclei.

1.5.4
Bone Marrow Cells

1.5.4.1
Unfractionated Bone Marrow

Unfractionated bone marrow cells have been reported to transdifferentiate into skeletal muscle, hepatocytes and neurons. Ferrari et al. (1998) reported that bone marrow cells of *lacZ* transgenic mice could differentiate into muscle cells in vivo when injected into the regenerating tibialis anterior muscle of *scid/bg* mice. β-Galactosidase-expressing cells appeared in regenerating muscle fibers, suggesting that the bone marrow cells augmented regeneration by resident satellite cells. However, the number of these transdifferentiated cells was very low (~0.2% of myofibers) and they appeared in the muscle only after a lag time of 2–5 weeks compared to injected satellite cells, which were incorporated into the regenerating muscle within 2–5 days after injection.

LaBarge and Blau (2002) reported transdifferentiation of bone marrow cells into satellite cells during irradiation-induced muscle regeneration. Bone marrow from mice transgenic for GFP were transplanted via tail vein into lethally irradiated syngeneic mice. Single isolated myofibers were isolated 2–6 months later and examined by confocal scanning microscopy in conjunction with immunohistochemistry. Diploid GFP[+] cells staining for the satellite cell markers cMet-R and α7 integrin were found in the satellite cell compartment of the myofibers and contributed in low numbers (<1% of the fibers) to regenerated muscle. Prolonged exercise-induced damage (on a running wheel) increased the number of myofibers expressing GFP$^+$ to 3.5% of the total. These results and those of Ferrari et al. (1998) fit with the evidence that muscle contains MuSCs that originate in the bone marrow and replenish the satellite cell population in injured muscle.

Sell (2001) has argued that the stem cells of the bile ductules that give rise to oval cells also have their origin in the bone marrow. Bone marrow reconstitution assays suggest that bone marrow cells can transdifferentiate into hepatic oval cells and hepatocytes after irradiation and liver damage (Petersen et al. 1999). Bone marrow from wild-type male mice was injected into lethally irradiated female mutant mice lacking the enzyme dipeptidyl peptidase IV (DPPIV). The recipients were treated with 2-AAF to prevent hepatocyte proliferation, then with CCl$_4$ to damage the liver, evoking an oval cell response. When the recipient livers were examined 2 weeks later, about 1×10^6 hepatocytes and oval cells were positive for the Y chromosome or for DPPIV. This number represents about 0.15% of the total hepatocytes and oval cells. In another experiment, 1%–2% of the hepatocytes of lethally irradiated female mice transplanted with male bone marrow were positive for the Y chromosome (Thiese et al. 2000a). Clinical studies of female patients with liver damage who received male bone marrow transplants also have identified Y$^+$ hepatocytes in the recipients at a frequency of 0.5%–2% (Alison et al. 2000; Theise et al. 2000b).

Brazelton et al. (2000) showed that GFP-expressing bone marrow cells injected into lethally irradiated hosts not only reconstituted the hematopoietic system, but also transdifferentiated into neurons of the olfactory bulb, hippocampus, cortex, and cerebellum. Confocal microscopy was used to show that GFP and neuronal-specific proteins (NeuN, NF-H) were expressed in the same cells and that these cells displayed neuronal morphology. An estimated 0.2%–0.3% of the total number of neurons in the olfactory bulb were derived from bone marrow cells. In a different kind of assay, newborn female PU.1-null mice received intraperitoneal transplants of wild-type male bone marrow cells (Mezey et al. 2000). PU.1 is a transcription factor expressed in hematopoietic cell lineages. Mice lacking this factor lack immune cells and, without a bone marrow transplant, die within 48 h after birth. Between 1 and 4 months after receiving the male bone marrow, the recipients were probed for Y$^+$ cells that also reacted to antibodies against NeuN and NSE. From 2.3% to 4.6% of all cells in the brain were Y$^+$, as were 0.3%–2.3% of cells that were NeuN$^+$ and NSE$^+$. The latter cells were found in the hypothalamus, hippocampus, striatum, amygdala, and especially the cortex.

1.5.4.2
Hematopoietic Stem Cells

Several reports have suggested that HSCs have a wide developmental potential. FACS-purified HSCs of male mice with the KTLS phenotype [Sca-1, c-kit, CD43, CD45]$^+$ [Lin CD34]$^-$ were injected into lethally irradiated *mdx* female mice. The tibialis anterior muscle was examined 12 weeks later for Y$^+$ cells that expressed dystrophin (Gussoni et al. 1999). Up to 4% of the muscle fiber nuclei were positive for dystrophin and the Y chromosome, suggesting that HSCs can transdifferentiate into muscle cells during the ongoing regeneration characteristic of *mdx* mice. Analysis of the cardiotoxin-injured tibialis anterior muscles of lethally irradiated C57B1/6 mice that received ROSA26 HSCs revealed the presence of β-galactosidase-expressing cells in the regenerated muscle (Castro et al. 2002). However, ROSA26 HSCs failed to differentiate into neurons in irradiated C57B1/6 hosts when injected either before or after cortical stab injury (Castro et al. 2002). A few β-gal-positive cells were observed, but these were associated with blood vessels and did not have the morphology of neurons. This would suggest that the bone marrow cell type that formed neurons after injection of unfractionated bone marrow is not the HSC.

Krause et al. (2001) used elutriation to isolate a fraction of mouse male bone marrow enriched in HSCs. The cells of this fraction were labeled with the membrane dye PKH26 and injected into irradiated females. The PKH26+ cells homed to the bone marrow and were recovered from the recipients by flow cytometry. The cells were diluted to one-per-unit volume and injected into 30 lethally irradiated females. Five mice survived, indicating reconstitution of their hematopoietic system. Donor Y+ cells were found in the epithelia of the respiratory and digestive systems, and the epidermis of the skin, at frequencies from 0.19% in the large intestine to 20.3% in the alveoli of the lungs. The differential frequency of apparent transdifferentiation to epithelial cells is related to the extent of irradiation damage suffered. Lung epithelium is more easily damaged compared to intestinal epithelium, which might account for the high frequency of donor cells found in the alveoli.

1.5.4.3
Mesenchymal Stem Cells

Mesenchymal stem cells normally differentiate into chondrocytes, osteocytes, and adipocytes in vitro (Friedenstein et al. 1968; Owen and Friedenstein 1988; Pittenger et al. 1998). They can be separated from HSCs in culture by the fact that they are adherent to the culture dish, whereas HSCs are not. Human MSCs are positive for the antigens SH2, SH3, CD29, CD44, CD71, CD90, CD106, CD120a, and CD124, and negative for the hematopoietic markers CD14, CD344, and CD45 (Pittenger et al. 1998).

The ability of MSCs to differentiate into other cell types has been tested in vitro and in vivo. MSCs have been directed to differentiate in vitro into skeletal muscle, cardiac muscle, and neurons. Treatment of mouse MSCs with 5-azacytidine in vitro induces them to differentiate as skeletal muscle cells (Wakatani et al. 1995). At least two MSC lines can differentiate as beating cardiomyocytes with a contractile protein profile of fetal ventricular cardiomyocytes, either spontaneously after long-term culture (Jiang et al. 2000) or after 5-azacytidine treatment (Makino et al. 1999). Both these lines express the cardiac-specific transcription factor Nkx2.5.

Adult rat and human MSCs have been induced by β-mercaptoethanol or dimethyl sulfoxide/butylated hydroxyanisole treatment to differentiate at high frequency (up to 80%) in vitro into neural-like phenotypes expressing the neural markers NSE, NeuN, neurofilament-M, and tau (Woodbury et al. 2000). Individual clones of cells were self-renewing, giving rise to both neurons and MSCs. In another set of experiments, human or mouse MSCs cultured in neural differentiation medium containing *all–trans* retinoic acid and BDNF differentiated at a low frequency into neural (0.5%) or glial (1%) phenotypes expressing NeuN or GFAP (Sanchez-Ramos et al. 2000). Human MSCs labeled with red or green fluorescent tracker dyes or mouse MSCs transgenic for *lacZ* were shown to express NeuN at a frequency of 2%–5% or GFAP at a frequency of 5%–8% when co-cultured with mouse fetal midbrain cells. The morphology of these cells approximated that of immature neural and glial cells (Sanchez-Ramos et al. 2000).

Assays in vivo suggest that MSCs can transdifferentiate into a variety of cell types. Cultured rat MSCs labeled with BrdU were induced to differentiate as cardiomyocytes when transplanted into myocardium in vivo (Tomita et al. 1999; Bittira et al. 2000; Wang et al. 2000). Marked MSCs infused into X-irradiated mice homed to the bone marrow and were later found in the connective tissues of liver, lung, and thymus parenchyma at a frequency of 0.2%–2.3% (Pereira et al. 1995; Prockop 1997). When injected into the brain ventricles of neonatal mice, MSCs migrated throughout the forebrain and cerebellum, where they differentiated into astrocytes (Azizi et al. 1998; Kopen et al. 1999).

1.5.4.4
Multipotential Adult Progenitor Cells of Bone Marrow

Recently, a multipotential cell has been isolated from human, rat, and mouse marrow cultures that have undergone 15 or more doublings (Reyes et al. 2001, 2002; Schwartz et al. 2002; Jiang et al. 2002). These cells, designated multipotent adult progenitor cells (MAPCs) are smaller (8–10 μm diameter) than MSCs and have the antigenic phenotype [CD34, CD44, CD45, c-kit, MHC complex class I and II]$^-$. They express low levels of Flk-1, Sca-1, and Thy-1, and higher levels of CD13. The cells show similarities to embryonic stem cells (ESCs) in their expression of stage-specific antigen I (SSEA-I) and the transcription factors Oct-4 and Rex-1, and in the dependence of the murine cells and independence of the human cells, on LIF to maintain the undifferentiated state. They express high levels of telomerase and their average telomere length has remained unchanged for over 100 population doublings (Reyes et al. 2001; Jiang et al. 2002).

In vitro, clones of single GFP-labeled MAPCs have been induced to differentiate with the morphology, antigenic phenotypes, and functional characteristics of endothelial cells, hepatocytes, and neural cells. In vivo, when ROSA-26 MAPCs were assayed for developmental potency in the chimeric embryo assay (C57BL/6 hosts) after 55–65 population doublings, chimerism was detected in 80% of the mice derived from blastocysts receiving 10–12 cells and in 33% of those receiving one cell. In both cases, the extent of chimerism in individual animals ranged between 0.1% and 45%. Sections of whole chimeric mice and individual organs were examined for β-galactosidase staining along with phenotype-specific markers. The MAPCs contributed to virtually every tissue of the chimeric mice: brain (including cortex, striatum, hippocampus, thalamus, and cerebellum), retina, lung, myocardium, skeletal muscle, liver, intestine, kidney, spleen, bone marrow, blood, and skin.

1.5.5
What Is the Basis of Adult Stem Cell Potency?

The developmental potency of ASCs that has been observed may or may not be due to transdifferentiation (Fuchs and Segre 2000; Weismann 2000). Some of the observed results have been shown to be due to cell fusion to form heterokaryons in which the transcriptional profiles of the partner cell are adopted by the ASCs through intracellular reprogramming, as opposed to responding to extracellular signals. This type of reprogramming by heterokaryon formation between differentiated cell types has been demonstrated many times in vitro (Blau et al. 1985) and would be particularly plausible in the case of cells that appear to become muscle.

The first demonstration that cell fusion might be responsible for purported cases of transdifferentiation came from experiments in which GFP-labeled mouse NSCs or bone marrow cells were cultured with ESCs to determine whether they could be converted to the ESC phenotype. GFP-labeled cells that expressed ESC markers and were LIF-dependent for suppression of differentiation appeared in the cultures. These cells could form embryoid bodies that differentiated spontaneously in the absence of LIF into several different cell types. However, karyotyping revealed that the cells had a 4 N chromosome number and exhibited heterochromatins of both ASCs and ESCs (Terada et al. 2002; Ying et al. 2002). The frequency of fusion in these experiments was calculated at 10^{-6}–10^{-4}.

Fusion at high frequency has now been demonstrated in the in vivo rescue of the liver of tyrosinemic (FAH$^-$) mice by FACS purified HSCs transplanted at limiting dilution. Such transplants had appeared to give rise to cells morphologically similar to hepatocytes that extensively reconstituted the liver (Lagasse et al. 2000). However, a repeat of this experiment showed that the new liver cells were positive for molecular markers of both donor and host (Vassilopoulos et al. 2003; Wang et al. 2003). Interestingly, donor cell markers were seen only after the host hematopoietic system has been reconstituted. This suggests that it is not the HSCs themselves that are fusing with host hepatocytes, but differentiated products of the HSCs such as macrophages, which fuse normally to form osteoclasts.

The frequency of fusion in the FAH$^-$ hosts was as high as that observed for the adoption of out-of-lineage phenotypes by neurospheres or MAPCs in chimeric embryo assays. Phenotypic conversion in bone marrow and in vitro assays is usually 2%–4%, though the frequency for MAPCs was as high as 13% in lightly irradiated or *scid* hosts (Jiang et al. 2002). Tests for fusion have not been carried out in other experiments, but the results of Wang et al. (2003) and Vassilopoulos et al. (2003) suggest that donor/host cell fusion could be a major factor in interpretation of the results of experiments designed to test for transdifferentiation and that such tests need to be carried out.

MAPCs and other stem cells isolated in vitro have been induced to differentiate into a number of foreign cell phenotypes, suggesting that transdifferentiation can indeed occur. However, even this is suspect, because it is possible that a rare, primitive pluripotent stem cell, similar to the pluripotent ESC, persists throughout development. This might be the cell that proliferates and differentiates into a wide variety of cell types when challenged in various assays (Fig. 2). Its prospective significance would be equal to its prospective potency. The existence of such a cell would be of great interest, since it would be a universal adult stem cell that could be used to regenerate any tissue.

What might be the origin of such a cell? One hypothesis is that the bone marrow might be a central reservoir for a pluripotent stem cell that gives rise not only to HSCs and MSCs in the marrow, but also circulates to give rise to pluripotent and multipotent adult stem cell populations of all tissues (Owen and Friedenstein 1988; Caplan 1991). Several lines of evidence are consistent with this idea. First is the fact that muscle has been demonstrated to contain MuSCs that can give rise to satellite cells and hematopoietic cells and that MuSCs originate in the bone marrow (Seale and Rudnicki 2000; Seale et al. 2000; McKinney-Freeman et al. 2002). Second, hepatic oval cells, HSCs, and MuSCs share common surface antigens and developmental properties. Third, Young and colleagues (1999) have reported the isolation of pluripotent and multipotent MSCs from the connective tissue compartments of adult rat and human muscle and from granulation tissue. The human cells are positive for CD10, CD13, CD56, and MHC class I antigens (Young et al. 1999). Fourth, Vacanti et al. (2001) have described "spore cells" residing in a variety of tissues that can differentiate into the functional cell type of the tissue of residence. Light, scanning, and transmission electron microscopy showed these cells to be small, rounded entities, 3–5 μm in diameter, with a large nucleus and scant cytoplasm. BrdU incorporation studies indicated that the cells have a doubling time ranging from 12–36 h, depending on the tissue of derivation. When cultured in a DMEM/F-12 medium, the cells reproduced tissue-specific morphology after 7–10 days in culture. Retina-derived cells developed into larger structures with the morphology of rods and cones, cells from spinal cord differentiated into neurons with connections resembling synapses, pancreas-derived cells formed islets that stained positively with a monoclonal antibody to insulin, liver-derived cells differentiated into bile-containing cells with hepatocyte morphology, and lung-derived cells developed a branching morphology. The spore cells were reported to remain viable after freezing tissues at −86°C for 30 min.

Another possibility is that a pluripotent cell could be created by dedifferentiation of adult stem cells, reprogramming nuclei to be capable of responding to all the signals required for development (Fig. 2). Dedifferentiation is a distinct possibility when cells are cultured for many generations in vitro, as was the case for the isolation of MAPCs (Jiang et al. 2002). There is no reason to believe that pluripotency cannot be achieved in this way, given the fact that reprogramming given of somatic nuclei has been demonstrated many times (Wilmut et al. 2002).

Some experiments testing the developmental potency of HSCs or NSCs by the bone marrow reconstitution assay have been unable to confirm the developmental plasticity observed by others. Wagers et al. (2002) isolated KTLS HSCs from the bone marrow of GFP-transgenic mice by FACS. Surviving animals were analyzed 4–9 months later for the transdifferentiation of GFP+ cells by staining with antibody to the pan-hematopoietic marker CD45 and with tissue-specific antibodies, coupled with standard and confocal microscopy. One GFP+ Purkinje neuron was found out of over 13 million cells examined in the brain, and only 7 out of 470,000 hepatocytes examined were GFP+. No transdifferentiated cells were found in kidney, gut, skeletal muscle, cardiac muscle, or lung. These low levels of GFP+ differentiated cells could easily be explained by the fusion of HSCs to neurons or hepatocytes. Castro et al. (2002) reported that neither unfractionated ROSA26 mouse bone marrow cells or side population cells of bone marrow transdifferentiate into neural cells in vivo, though side population cells can differentiate into other cell types such as cardiomyocytes and endothelial cells (Goodell et al. 1996, 1997; Jackson et al. 2000). In another experiment, clonally derived NSCs injected into irradiated mice failed to reconstitute the bone marrow and did not regenerate any blood cells (Morshead et al. 2002). All these negative results might be because the cells tested did not include a pluripotent stem cell or did not allow for dedifferentiation to pluripotency. Part of the problem in assessing whether ASCs are truly multipotent or pluripotent as they exist within tissues in vivo is the many differences in experimental protocols among experiments. To be certain about what is going on, we need to conduct experiments according to protocols that standardize the isolation and purification of ASCs, labeling, culture times and conditions, assay systems, differentiation markers, microscopy, and checks for cell fusion.

Whatever the explanation of the developmental potential observed in these experiments and its relevance to normal physiology, such potential has enormous significance for regenerative medicine because ASCs could be expanded in vitro for use as autogeneic cell transplants to replace tissues damaged by injury or disease. The issue is then whether injury environments contain the minimum number of developmental signals required to make a stem cell transdifferentiate. The experimental evidence from bone marrow reconstitution assays in irradiated and *scid* mice would suggest that they do. Most tissue-specific injury environments, however, have not been tested for their ability to support the developmental potential of multipotent or pluripotent cells. Furthermore, although morphology, molecular markers, and electrical and biochemical function tests indicate that the differentiated donor cells are the real thing, it is not clear what their level of function is or how long they remain functional, since behavioral and longevity studies of host animals have not been reported.

2 Regenerative Medicine

Investigations over the past two decades have established the feasibility and promise of translating findings from regenerative biology into a regenerative medicine. The majority of this work has been done with experimental animals, but a few human clinical trials have been performed as well. Two basic strategies have been used: cell transplants and the induction of regeneration in vivo (Fig. 3).

2.1
Cell Transplants

The cell transplant approach involves replacing the cells of a damaged tissue by transplanting autogeneic, allogeneic, or xenogeneic cells into the lesion. The cells can be transplanted as suspensions or aggregates, or embedded in a matrix to construct a bioartificial tissue ("tissue engineering"). The cells can be differentiated cells, fetal cells, adult stem cells, or precursor cells derived from embryonic stem cells (ESCs); they can be normal or genetically modified to boost production of important signaling or ECM molecules, or molecules that neutralize factors inhibitory to cell survival and proliferation. The advantage of allogeneic or xenogeneic cells is that they can be

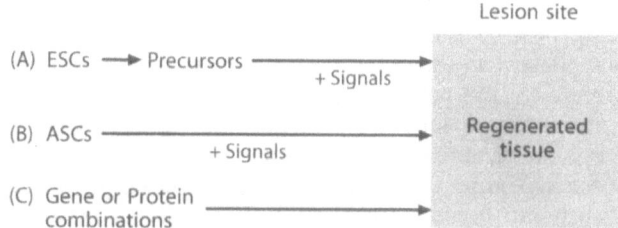

Fig. 3A–C Strategies of regenerative medicine. **A** Embryonic stem cells are directed to differentiate in vitro into precursors of the desired cell type. The precursors are transplanted into the lesion as suspensions, aggregates, or in a biomimetic scaffold. If necessary, signaling factors such as growth factors, are incorporated into the transplant. The cells will respond to these signals, or to signals endogenous to the injury environment, by regenerating new tissue. **B** Adult stem cells are transplanted into the lesion site, with or without exogenous signaling molecules, where they will respond by regenerating new tissue. **C** Chemical induction of regeneration by incorporation of regeneration-promoting genes into tissues of the lesion or delivery of regeneration-permissive proteins to the lesion site

cultured and banked in advance for use in cases where there is not enough time to expand a patient's own cells for use; the disadvantage is that they will elicit rejection if seen by the immune system. Most studies have focused on adult stem cells and derivatives of embryonic stem cells.

A major advantage of ESCs is that they can be expanded indefinitely in culture without losing their potential for differentiation. However, it is not easy to direct the differentiation of ESCs and cells derived from them are allogeneic and therefore subject to immune rejection. Furthermore, the use of ESCs and fetal cells is surrounded by bioethical controversy, because their production requires the destruction of a human embryo. Adult stem cells can be harvested from the patient and therefore have the advantage of not provoking an immune response or generating ethical controversy. A potential disadvantage is the difficulty of harvesting and expanding many types of ASCs. Bone marrow cells, however, are relatively easy to harvest and expand. The use of both ESCs and ASCs for therapeutic purposes is currently the subject of research in many laboratories around the world (Fuchs and Segre 2000; Weissman 2000; Blau et al. 2001).

Ideally, the matrices of bioartificial tissues would mimic the ECM in vivo, providing not only the geometry and physical/chemical properties to maximize the migration of cells, but also capable of sequestering and releasing biological signals essential for proliferation and differentiation. In addition, scaffolds should incorporate factors that neutralize molecules inhibitory to regeneration. Collagen I is the predominant matrix protein in the body and is the biomaterial most widely used for bioartificial tissue construction, either by itself, or in combination with other ECM molecules. It is easy to extract and mold into a variety of shapes, and is biodegradable. Other frequently used scaffold materials are alginate, polydioxanone, poly (epsilon-caprolactone), poly (glycolic acid), and poly (lactic acid). All except alginate (in crosslinked form) are biodegradable.

Bioartificial tissues can be open (vascularized by the host) or closed (cells encapsulated and dependent on diffusion for survival). Closed constructs require the use of already differentiated cells; those of open constructs could be ASCs or precursors derived from ESCs. If the cells of an open construct are allogeneic or xenogeneic, they will be subject to immunorejection; those of a closed construct will not. The matrix supporting a closed construct must be resistant to degradation. By contrast, scaffolds of open constructs should be biodegradable so that they are replaced over weeks and months with the natural matrix made by the cells. Other challenges to building open constructs include the incorporation of angiogenic factors into the scaffold of open bioartificial tissues and how to create three-dimensional patterns of cells that allow for maximum diffusion while blood vessels grow into the proliferating tissue, or incorporate blood vessels into the tissue as it is built.

Cell transplants have been used as experimental therapies in both animals and humans to treat a wide variety of diseases and injuries. Several of these will be discussed by tissue and organ system.

2.1.1
Central Nervous System

2.1.1.1
Demyelinating Disorders

Brustle et al. (1999) used glial cell precursors differentiated in vitro from mouse ESCs to treat a rat model of a human demyelinating disorder, Pelizaeus-Merzbacher disease. This disease is caused by a mutation in the X-linked gene for myelin proteolipid protein (PLP). The mouse ESCs were placed in defined medium with FGF-2 and PDGF to produce proliferating glial precursors. These precursor cells were then injected into the spinal cord and brain of 7-day-old myelin-deficient rats. Two weeks later, electron microscopy and staining of the tissues with a probe to mouse satellite DNA and antibody to PLP showed that the injected cells had differentiated into oligodendrocytes that formed myelin sheaths around the axons.

Human olfactory ensheathing cells were used to remyelinate dorsal spinal cord axons in rats that were focally demyelinated by X-irradiation and injection of ethidium bromide (Kato et al. 2000). The olfactory ensheathing cells, prepared from adult olfactory nerves removed during surgery for nasal melanoma, were injected into the lesioned areas. Examination of tissue sections 3 weeks later by electron microscopy and in situ hybridization with the COT-1 human DNA probe revealed remyelination of the axons by the human cells.

Multiple sclerosis (MS) is an autoimmune CNS disorder in which myelin is damaged and astrocytes proliferate to form scar. The result is blockade of electrical impulses along nerve axons, leading to loss of sensation and coordination, paralysis, and blindness. Experimental work on a mouse model of MS, autoimmune encephalomyelitis (EAE) and clinical trials in humans suggest that attacks of MS can be aborted or diminished by blocking the $\alpha4$ integrin molecule on the surface of the attacking immune cells, or by the use of agents that reduce the production of inflammatory cytokines and metalloproteases by inflammatory cells (Yednock et al. 1992; Steinman 2001; Miller et al. 2003). Other methods under development to suppress autoimmunity could also be useful in halting the progression of MS, but regardless of how far the disease has progressed, any existing damage needs to be repaired. This has been accomplished in EAE mice by the intravenous injection of neurospheres derived from NSCs of the lateral ventricles of the brain (Pluchino et al. 2003). The NSCs express the same $\alpha4$ integrin expressed by attacking immune cells; they homed to sites of demyelination where they differentiated into oligodendrocytes and new neurons. Astrogliosis and axon damage were markedly reduced. Remyelination with complete functional recovery from paralysis was the outcome in 26.6% of the mice, whereas controls showed no sign of recovery. Interestingly, donor NSCs provided only 20% of the new oligodendrocyte precursors in sites of remyelination, the remainder being host cells. This result suggests that the injected cells can regulate the behavior of host oligodendrocytes and astrocytes.

2.1.1.2
Spinal Cord Injury

Partial restoration of function after contusion of rat spinal cords has been accomplished by injecting neural/glial precursors differentiated from mouse ESCs in vitro into the lesion 9 days after injury (McDonald et al. 1999). Staining with antibodies specific for mouse proteins and for glial and neuronal markers showed that many of the implanted cells survived, migrated throughout the injured area, and differentiated into new interneurons, oligodendrocytes, and astrocytes. Partial recovery of function was indicated by the ability of the rats to bear weight on their hind legs and by restoration of partly coordinated stepping movements.

The mechanism of recovery in these experiments is not clear. The mouse neurons may have made functional connections with host rat neurons, partly restoring signal transmission between brain and hind legs, or the mouse oligodendrocytes may have rebuilt myelin sheaths around demyelinated host axons, enabling them to conduct impulses again. It is also possible that the mouse cells secreted molecules that were neuroprotective to damaged host cells, preventing them from dying or restoring the ability of axons to regenerate.

2.1.1.3
Parkinson's Disease

Parkinson's is an invariably progressive disease characterized by tremors at rest, akinesia, and bradykinesia, muscle rigidity, postural instability and lack of facial expression. Normal movement is regulated by the striatopallidothalamic output pathway (SPTOP). The neurotransmitter dopamine, produced by cells in the substantia nigra of the striatum, maintains a normal level of output from the SPTOP to the motor cortex. Parkinson's is caused by the death of dopaminergic neurons (DANs) in the substantia nigra, leading to lower dopamine output, hyperactivity of the SPTOP, and impaired motor function.

There are three primary therapies for the disease (Bjorkland and Lindvall 2000). The first is to increase the dopamine output of the remaining viable DANs by administration of L-dopa, which is taken up and converted to extra dopamine that compensates for the lost cells. The second therapy is pallidotomy, and the third is inhibitory electrical stimulation of the subthalamic nucleus, which normally stimulates the globus pallidus. All these therapies reverse the akinesia, bradykinesia, and rigidity of the disease, but not the tremors. L-Dopa treatment produces severe side effects after several years and eventually has no effect on disease symptoms when the number of viable DANs drops too low. Many symptoms recur a few years after pallidotomy and electrical stimulation of the subthalamic nucleus. None of the therapies slow the disease progression. Thus, investigators have turned to transplants of DANs in the hope of a cure. These transplants have been performed on both human patients and a rat model of Parkinson's. In this model, the neurotoxin β-hydroxydopamine is injected unilaterally into the striatum. The rats suffer muscle rigidity on one side and exhibit turning movements, as well as akinesia and bradykinesia.

Fetal human mesencephalic cells (which include dopaminergic NSCs) differentiate into DANs and make synaptic connections with striatal neurons when injected into the striatum of Parkinson's patients, restoring the activity of the SPTOP toward normal. However, the results of such transplants have been highly variable. In the best cases, there have been dramatic clinical improvements that have lasted 5–10 years (Bjorkland and Lindvall 2000; Lindvall and Hagell 2001). These improvements are correlated with an increased output of dopamine, as visualized by increased uptake of ^{18}fluoro-dopa in positron emission tomography (PET) scans. In other cases, improvements have been minimal, or patients have continued to deteriorate. Autopsies of two patients who died, as well as transplant experiments on Parkinsonian rats indicate that this variation is due to differential survival of transplanted cells. It is thought that a minimum of 80,000 DANs (~20% of the normal number of DANs in the human substantia nigra) are required to obtain a beneficial effect.

Practical and ethical considerations dictate that fetal tissues cannot be a reliable source of cells for transplant; thus, other sources of DANS are being explored. Carotid body cells, which regulate respiratory rate through the medulla by monitoring changes in blood levels of O_2, CO_2, and H^+, are rich in dopamine and thrive in low O_2. Since only one carotid body is necessary, the other could be used as a source of DANS for autotransplants. When injected into the striatum of rats with Parkinson's, over 90% of the carotid body cells survived and dopamine levels increased to 65% of unlesioned controls (Espejo et al. 1998). The rat striatum is normally innervated by ~30,000 DANs, but only 400–600 carotid body cells were needed to achieve this result. Turning movements were eliminated, but there was less improvement in akinesia and bradykinesia.

Transplantation of ESC derivatives is also being explored as a treatment for Parkinson's. Kim et al. (2002) used a five-stage protocol (Lee et al. 2000) to differentiate mouse ESCs transfected with the nuclear receptor related-1 (nurr-1) gene into DANS. Nurr-1 is a transcription factor that promotes the differentiation of mesencephalic precursor cells into tyrosine hydroxylase (TH)$^+$ dopaminergic cells in the presence of FGF-8 and Shh (Hynes and Rosenthal 1999; Wurst and Bally-Cuif 2001). Grafts of 5×10^5 cells were injected into the striatum of rats lesioned unilaterally by β-hydroxydopamine. The ESC-derived TH$^+$ cells were functional DANS as assessed by morphological, neurochemical, electrophysiological, and behavioral criteria. They released dopamine and extended axons into the host striatum. The axons formed functional synapses as indicated by significant recovery from amphetamine-induced rotation and improvement in step-adjusting, cylinder, and paw-reaching tests. Since the brain is an immune-privileged site, the next step would be to determine whether DANs derived from human ESCs will lead to improvement of motor function in the rat Parkinson's model.

2.1.2
Myocardial Infarction

To establish "proof of concept" that grafted cardiomyocytes could integrate into host heart tissue, Soonpa et al. (1994) injected suspensions of fetal cardiomyocytes from *lacZ* transgenic mice into the uninjured ventricular myocardium of syngeneic host mice. The injected cells continued to proliferate and differentiated into mature cardiac muscle integrated with that of the host. Electron microscopic analysis indicated that the donor cells formed intercalated discs with the host myofibers, suggesting donor/host electrical coupling. No cardiac arrhythmias were noted. Similar results were obtained after transplanting fetal cardiomyocytes into the ventricular myocardium of dystrophic dogs (Koh et al. 1995). Cardiomyocytes differentiated from mouse ESCs in vitro were transplanted into the ventricular myocardium of of *mdx* dystrophic mice (Klug et al. 1995, 1996). Staining with antibodies to dystrophin showed that the transplanted cells were stably integrated into the host cardiac muscle.

Another approach has been to inject skeletal muscle myoblasts or satellite cells into uninjured or injured heart tissue. C2C12 myoblasts injected into the uninjured myocardium of syngeneic mice withdrew from the cell cycle and formed typical myotubes (Koh et al. 1993). No overt cardiac arrhythmias were noted and the donor cells survived for as long as 3 months. Data were not obtained as to whether the skeletal myotubes became electrically coupled to host cardiomyocytes or whether their contractile activity contributed to the function of the myocardium. Taylor et al. (1998) induced myocardial infarction in rabbit hearts by cryoinjury or coronary ligation. They injected 10^7 skeletal myoblasts into the infarcts and found that many of them differentiated into cells with the characteristics of cardiomyocytes connected by intercalated discs, while others differentiated into multinucleated skeletal muscle. Cardiac function was improved, but whether the mechanism of improvement was due to contraction of the muscle formed by the donor cells or to a lessening of mechanical stiffness of the scar by the presence of new muscle, or both, was not clear. Others have reported that satellite cells transplanted into cryo-infarcted ventricular muscle of rats, rabbits, and pigs integrated into the heart muscle, differentiated into cardiomyocytes and improved heart function (Chiu et al. 1995; Atkins et al. 2000a,b; Kao et al. 2000).

The first phase I trial transplanting satellite cells into the damaged human heart was carried out by Menasche (2000) on a 72-year-old patient suffering from severe congestive heart failure caused by extensive myocardial infarction. Satellite cells were isolated from a quadriceps muscle biopsy, expanded in vitro for 2 weeks and 800×10^6 cells (65% myoblasts) delivered into the myocardial scar via 30 injections with a small-gauge needle. Simultaneously, a double bypass was performed in viable but ischemic areas of the myocardium. Six months later, the patient's symptoms were dramatically improved. Echocardiogram showed evidence of new-onset contraction and fluorodeoxyglucose PET scan showed increased metabolic activity of the infarct. The improvement was considered unlikely to be due to increased collateralization from the bypass region, because this region was far from the infarct. Since this trial, several other cardiac patients have been transplanted with satellite cells.

The apparent plasticity of bone marrow cells has led several groups to use them in treating myocardial infarction. Orlic et al. (2001) injected 1×10^5 Lin$^-$ c-kit$^+$ HSCs from male mice transgenic for GFP into the viable myocardium of female hearts 3–5 h after infarction by coronary ligation. BrdU was administered each day for 4–5 days. Nine days after grafting, ventricular function was measured and sectioned hearts were immunostained for BrdU, GFP, Ki67, and cardiac-specific proteins. Hemodynamic function in the grafted animals improved by 40%. GFP$^+$ cells incorporated BrdU, were positive for Ki67, and expressed cardiomyocyte-specific proteins. The grafted HSCs appeared to migrate into the infarct area and form new myocardium and blood vessels. Closely packed myocytes occupied 68% of the infarct and connexin 43 expression was detected at the surface of closely aligned cells, suggesting the formation of gap junctions.

In a similar study, ROSA26 HSCs were used to reconstitute the marrow of lethally irradiated host mice, after which coronary ligation was performed (Jackson et al. 2001). Subsequent analysis revealed that, within the infarct, 0.02% of the cardiomyocytes, and up to 3.3% of the endothelial cells in small blood vessels, were donor-derived. However, little functional improvement was noted, perhaps because of the small number of new cells. Kocher et al. (2001) isolated hemangioblasts from human bone marrow, expanded them in vitro, and injected the cells into rat myocardial infarcts induced by coronary ligation. By 2 weeks, there was a significant increase in the microvascularity of the infarct, with human cells accounting for 20%–25% of the increase. After 15 weeks, the reduction in hemodynamic function of the experimental hearts had decreased to 18%–34%, compared to 48%–59% in controls.

All of these observations suggest that the HSCs, or a more primitive multipotent or pluripotent stem cell included in the transplants, responded to the injury environment by becoming cardiomyocytes and endothelial cells.

2.1.3
Osteogenesis Imperfecta

Whole bone marrow has been tested for its ability to alleviate the symptoms of osteogenesis imperfecta, or "brittle bone" disease. In this disorder, a mutation in the type I collagen gene makes the bone matrix highly susceptible to fracture, resulting in numerous hospitalizations. Horwitz et al. (1999) administered chemotherapy to three infants with osteogenesis imperfecta to destroy their immune systems, then transfused whole bone marrow from HLA-identical siblings to the patients. Bone biopsies revealed that only a small percentage of the donor cells engrafted, but in the 6 months following transplant, the number of fractures suffered was reduced by 92%. A possible next step would be to isolate MSCs from osteogenesis imperfecta patients, replace the mutated gene by homologous recombination in vitro, then return the cells to the patient.

2.1.4
Diabetes

Allotransplants of pancreatic islets from cadavers to treat diabetes were begun in 1990, but the harsh dissociation procedures and immunosuppressive regimes led to such poor cell survival that only 12.4% of recipients showed insulin-independence for more than 1 week and only 8.2% for more than 1 year. In a protocol now known as the Edmonton Procedure, Shapiro et al. (2000) introduced much gentler dissociation and immunosuppression methods that greatly enhanced donor cell survival. The donor islets are transplanted through a portal vein catheter into the liver. The procedure is done under local anesthetic and sedation and takes only 20 min. Two transplants several weeks apart, totaling between 65 and 200×10^6 β-cells, are required to attain normoglycemia, which occurs rapidly. The major drawback to this procedure, aside from the need for immunosuppression, is donor shortage; thus, stem cell transplants and bioartificial tissues are being investigated.

Islets generated in vitro from pancreatic ductal stem cells and encapsulated in hyaluronic acid reversed insulin-dependent diabetes when injected under the kidney capsule (Ramiya et al. 2000). A closed bioartificial pancreas using a non-biodegradable matrix was clinically tested in a 38-year-old male patient who had been diabetic for 30 years (Soon-Shiong et al. 1994). This construct consisted of high-guluronic acid alginate microcapsules containing islets harvested by collagenase digestion from 8 cadavers. The patient received a total dose of 15,000 encapsulated islets/kg body weight implanted into the intraperitoneal cavity in two doses of 10,000 and 5,000 islets 6 months apart. Three months after the second implantation, diabetic symptoms were markedly abated and the patient was able to discontinue insulin injections. Proinsulin levels, however, were high, suggesting that the dose of islets/kg should be increased. How long the structural integrity of the microcapsules can be maintained in patients and how long the islets can survive within the capsules are two important questions that have not yet been answered. Because of the shortage of human donor tissue, xenogeneic cells may be useful in making such microcapsules.

A hybrid plastic/cellular bioartificial pancreas has been tested in dogs (Maki et al. 1996). The device consists of a disc-shaped acrylic housing around a single-coiled membranous tube. The tube is attached to a vascular graft that forms an arteriovenous shunt between the iliac artery and vein. Pig islets are distributed around the coiled tube in an agar or alginate matrix and are nourished by blood circulating through the tube. The device remained patent for up to 3.5 years in normal dogs, demonstrating its excellent biocompatibility and was controlled the hyperglycemia of two pancreatectomized dogs for more than 8 months, as judged by normal levels of blood glucose and porcine C-peptide, low exogenous insulin requirement, and normal body weight. However, late vascular thrombosis was encountered in the majority of dogs used in the study.

2.1.5
Skin Injuries

Skin substitutes have been devised by a number of investigators (Boyce 1996; Yannas 2001) to treat damage to the dermis. Allogeneic dermal fibroblasts are seeded into an artificial matrix of polylactic or polyglycolic acid mesh or a matrix of collagen supplemented with other ECM molecules such as chondroitin sulfate and/or elastin. This artificial dermis is then overlaid with a meshed split-thickness skin graft (MSTSG) or cultured allogeneic keratinocytes and applied to a lesion. The removal of antigen-presenting dendritic cells ensures that the cells are only slowly rejected as host fibroblasts and keratinocytes replace those of the construct (Guerret et al. 2003). The regenerated skin is not perfect, however, and lacks completely normal histological structure, hair, and glands. Several commercial skin equivalents are in clinical use to treat burns, excisional wounds, and chronic venous stasis ulcers.

2.1.6
Cartilage and Bone Injuries

Autogeneic stem cells have been used to make bioartificial cartilage and bone (Dennis et al. 2001). Bone marrow-derived MSCs were seeded into ceramic matrixes and the construct was implanted to bridge gaps made in adult rat femurs (Caplan et al. 1993; Wakitani et al. 1994; Kadiyala et al. 1997). The MSCs differentiated into osteoblasts, probably through the action of growth factors in the host marrow or secreted by osteoblasts migrating into the ceramic from the periosteum and endosteum, and then into osteocytes that secreted new bone matrix as the ceramic was degraded by osteoclasts. Multipotential cells isolated from rabbit skeletal muscle have been used to repair defects in the articular cartilage of rabbit knee joints (Grande et al. 1995). The cells were seeded into porous polyglycolic acid matrixes and the constructs implanted into the defects. By 12 weeks postimplantation, the matrix had dissolved and the cells had formed a well-integrated surface layer of cartilage in the defect that was approximately the same thickness as the normal articular cartilage. The biochemical similarity of the bioartificial cartilage to normal cartilage, and its durability, is unknown.

2.1.7
Research Issues in Cell Transplantation

2.1.7.1
Cell Sources

Providing reliable sources of cells for cell transplants and bioartificial tissue construction requires either the expansion of freshly isolated autogeneic cells or the establishment of culture banks of allogeneic cells that can be drawn upon as required. It is also crucial to have culture media available that support the proliferation and differentiation of these cells. Many types of differentiated cells, such as hepatocytes and pancreatic islet cells, are difficult to expand in vitro because not all the factors essential for their proliferation and differentiation are understood. Thus, research on the factors that control the entry of stem, progenitor, and differentiated cells into, and progression through, the cell cycle, and the factors that control the differentiation of specific cell phenotypes are of prime importance for the design of media that allow cell expansion and differentiation (Parentau 2001). Some progress has been made in this direction; for example, a defined medium containing hepatocyte growth factor (HGF), epidermal growth factor (EGF), and TGF-β1 has been found to support the proliferation of mature rat and human hepatocytes (Block et al. 1996).

The pluripotency and ability of ESCs to proliferate indefinitely make them very attractive as a virtually unlimited source of cells for transplant and bioartificial tissue construction. The three main biological limitations to their use are the lack of protocols to direct their differentiation to the desired cell types, loss of imprinting and immunorejection of the differentiated cells when placed in non-privileged sites. Directed differentiation is possible for some cell types, however, and will become easier as we learn more about the signals that control cell differentiation during development. Differentiated cells must be derived from ESCs that have imprints of maternal and paternal alleles (one allele silenced, the other expressed). Injection of ESCs in which imprinting has not yet been established or has been lost, is associated with developmental abnormalities (Humpherys et al. 2001). The problem of immunorejection is theoretically solvable by therapeutic cloning. In this technique, blastocysts would be created by somatic nuclear transfer and used to create "personal" embryonic stem cell lines. Because they express the antigens of the nuclear donor, the differentiated progeny of these cells would not be immunorejected. Such lines have been established for mice (Wakayama et al. 2001) and proof of concept has been established for human cell lines (Cibelli et al. 2001). However, bioethical issues associated with the creation and destruction of blastocysts constitute an obstacle to the use of this technique in humans. A possible alternative is to derive ESC lines from blastocysts that have been created by suppression of polar body formation after artificial egg activation (Cibelli et al. 2002). These blastocysts cannot be brought to term because they are unable to implant in the uterus, but can be used to make ESCs.

Other strategies to avoid immune rejection of cells are (1) the creation of transgenic animals that express regulatory molecules that inhibit the activation or synthesis of key proteins involved in rejection; (2) mimicking immunoprivileged sites, such as the anterior chamber of the eye; (3) the production of antibodies directed at T cell receptors that recognize foreign cell surface multiple histocompatibility antigens (MHCs); and (4) the production of genetically modified "stealth" cells that do not express surface MHC antigens (Cooper and Lanza 2000).

2.1.7.2
Biomaterials for Bioartificial Tissues

Advances in biomaterials design are essential for the construction of bioartificial tissues and for stimulation of regeneration in vivo (Peppas and Langer 1994; Baldwin and Saltzman 1996; Griffith 2000). Such materials must meet several criteria. They must be "smart"; i.e., contain and/or be able to release appropriate biological signaling information to promote and maintain cell adhesion, differentiation, and tissue organization (Griesler 2001). In essence, we want to develop polymers that mimic the functions of the embryonic or adult ECM. Natural ECM consists primarily of type I collagen and fibronectin plus smaller amounts of other collagens and adhesion proteins, GAGs and proteoglycans, and tiny, but biologically potent, amounts of growth factors bound to ECM components (Badylak 2002). Thus, research on polymer chemistry must be coupled with research on the molecular biology of ECM signaling systems that regulate cell morphology and function and how to incorporate biological molecules into synthetic polymer matrixes. For example, selective cell adhesion to substrates could be achieved by linking specific oligopeptide or carbohydrate recognition sequences to non-adhesive artificial polymers. Such adhesive selectivity would be desirable, for example, in artificial blood vessels, where the materials used should support the adhesion and migration of endothelial cells but not the adhesion of platelets, which could trigger thrombosis. Growth factors might also be bound to polymer matrices and released as the polymers degrade (Saltzman 1996). If the matrix materials form a closed bioartificial system, they must allow the ready exchange of oxygen, nutrients, waste products, and molecules that regulate the metabolic and synthetic activities of the cells, while immunoprotecting cells. Other important properties of biomaterials are directional alignment to orient cells by contact guidance in particular directions essential for the function of the tissue (Yannas 2001), and controlled biodegradability (Griffith 2001).

2.2
Chemical Induction of Regeneration In Vivo

This strategy involves the use of combinations of regeneration-promoting molecules and neutralizers of regeneration-inhibiting molecules to stimulate regeneration-competent cells of the body to regenerate new tissue. The regeneration-competent cells might be fully differentiated cells induced to undergo compensatory hyperplasia, ASCs that have the potential to engage in regeneration but form scar tissue in the absence of intervention, or stem cells created by stimulating the dedifferentiation of mature cells. The regeneration-promoting molecules might be delivered as soluble "molecular cocktails" or as part of a cell-free regeneration template derived from tissue ECM or manufactured in the laboratory.

The advantage of this approach is that it eliminates problems of donor availability, cell culture, immunorejection, and bioethical concerns in one stroke, and would be relatively low-cost. The feasibility of inducing regeneration from the body's own tissues depends in part on whether non-regenerating tissues have a hidden capacity for regeneration, but are prevented from doing so by a fibrosis-promoting environment. There is substantial evidence that many, if not most, non-regenerating tissues harbor such latent capacity. Some of this evidence will be described here, along with attempts that have been made to chemically induce regeneration in vivo.

2.2.1
Evidence for Latent Regenerative Capacity in Mammals

2.2.1.1
New Cardiomyocytes Appear After Myocardial Infarction

In contrast to skeletal muscle, mammalian cardiac muscle does not regenerate (Polezhaev 1972). Myocardial infarcts are infiltrated by fibroblasts that form a scar and the remaining myocardium undergoes compensatory hypertrophy. Nevertheless, heart muscle responds to functional overload by DNA synthesis and mitosis. Significant increases in nuclear ploidy levels are observed in hypertrophied human and rat hearts subjected to prolonged functional overload or insufficiency (Oberpriller et al. 1983; Rumyantsev 1991; Borisov 1998, 1999). In rats, 4% of atrial and 0.1% of ventricular myocytes synthesize DNA after myocardial infarction but less than 2% of atrial and virtually no ventricular myocytes undergo mitosis (Borisov 1998). The ventricular myocardium of adult newts, however, is regenerated by proliferation of cardiomyocytes (Bader and Oberpriller 1978).

The assumption has been that human cardiomyocytes do not divide after myocardial infarction, but recent data suggest otherwise (Beltrami et al. 2001; Anversa and

Nadal-Ginard 2002). The expression of the mitosis-associated protein Ki-67 was examined in the hearts of 13 patients who died 4–12 days after suffering left ventricular infarction and in 10 control patients who died of other causes and had no major risk factors for heart disease. Nuclei were evaluated in the border zone around the infarct and at a more distant site. About 4% of the nuclei in the border zone expressed Ki-67 and 1% outside the border zone, amounting to nearly 2×10^6 cardiomyocytes in mitosis. This was 84 and 28 times the control values for these areas. The normal left ventricle contains about 5.5×10^9 cardiomyocytes, of which 1.7×10^9 were lost by infarct. Theoretically, these could have been replaced by mitosis in 18 days, assuming that mitosis lasts for 30 min and that myocytes divide only once. However, the products of division do not migrate into the infarct to replace the lost cells and it is repaired by scar tissue. The dividing cardiomyocytes appear to contribute to the hypertrophy of the remaining left ventricular muscle.

The simplest interpretation of these results is that the dividing cells are resident cardiomyocytes.

Like the cardiomyocytes of injured newt ventricle, mammalian cardiomyocytes can undergo partial dedifferentiation. Eppenberger et al. (1988) showed that mammalian cardiomyocytes are able to partially dedifferentiate in vitro. A high percentage (up to 25% ventricular and 63% atrial) of cultured mammalian cardiomyocytes synthesize DNA (Cantin et al. 1981, 1983; Claycomb 1991; Nag 1991), though only a very small percentage of (atrial) cells divide. Entry into the cell cycle and dedifferentiation appear to be linked in cardiomyocytes. The E2F-1 transcription factor transfected into cultured ventricular cardiomyocytes induces DNA synthesis while simultaneously inhibiting the cardiac and skeletal α-actin promoters, sarcomeric actin synthesis, and serum response factor, which is crucial for the transcription of both sarcomeric α-actins in cardiac muscle (Kirshenbaum et al. 1996). The ability to partially dedifferentiate, which is not shared by mammalian skeletal muscle, may be related to a difference in the relationship between proliferation and differentiation in developing skeletal and cardiac muscle. Skeletal myoblasts first proliferate, then withdraw from the cell cycle and fuse to form multinucleated myotubes. Just before fusion, the myoblasts express the transcription factors MyoD1 and myogenin (Claycomb 1992). These proteins function as a proliferation/differentiation switch that induces the expression of muscle-specific genes such as myosin and creatine kinase, while simultaneously inducing withdrawal of the myoblasts from the cell cycle. By contrast, the MyoD1 and myogenin genes are not expressed in cardiac myoblasts, and cardiac myoblasts proliferate and differentiate simultaneously (Claycomb 1992). Thus, cardiac myoblasts seem to have parallel programs for proliferation and cytodifferentiation, rather than serial programs linked by a single switch. This unusual situation is associated with the fact that the heart is one of the first organs to form during development and must become functional while still growing.

However, other interpretations of these results have not been ruled out. The heart might harbor stem cells that respond to injury by dividing and differentiating into cardiomyocytes, though no such cells have yet been identified. Or, stem cells might be recruited from other sources in the body, such as the bone marrow, and differentiate into cardiomyocytes. In support of this notion, Quiani et al. (2002) reported that up to 10% of the cardiomyocytes of female human hearts transplanted to male recipients were positive for the Y chromosome, indicating recruitment of cells from a host source.

2.2.1.2
Stem Cells Reside in Non-regenerating Tissues

In addition to the hippocampus and connective tissue compartments described earlier, stem cells have also been found in the spinal cord and neural retina. Functional recovery of the transected mouse spinal cord can occur if the transection is made in such a way as to minimize injury to the dura, fibroblastic infiltration, and displacement of the cut ends of the cord, all of which minimize fibrosis (Seitz et al. 2002). Neural stem cells reside in the ependymal layer and subventricular zone of the spinal cord (Weiss et al. 1996; Kehl et al. 1997; Johansson et al. 1999; Doetsch et al. 1999; Yamamoto et al. 2001a,b). When the cord is injured in vivo, ependymal NSCs divide to form an amplifying population of stem cells in the subventricular zone, which subsequently differentiate as reactive astrocytes that form glial scar tissue. However, when removed from the cord and cultured in vitro, these cells proliferate and differentiate into cells that have the morphological, immunocytochemical, and electrophysiological properties of neurons and glia (Weiss et al. 1996; Chiang et al. 1996; Kehl et al. 1997, Zhou and Chiang 1998; Johansson et al. 1999; Zhou et al. 2000).

Mammalian retinal neurons and thus optic nerve axons normally do not regenerate. However, dissociated rat retinal ganglion cells can regenerate processes in vitro on substrates of polylysine and collagen (Leifer et al. 1984), or on substrates soaked in medium conditioned by fish optic nerve (Schwartz et al. 1985). Stem cells capable of differentiating in vitro into neural retina-specific cell types have been identified in the pigmented ciliary margin of the mouse eye (Ahmad et al. 2000; Tropepe et al. 2000; Shatos et al. 2001). In addition, dissociated human pigmented epithelial cell lines from the eye of an 80-year-old donor have been shown to have the capacity to transdetermine in vitro into neuronal and lens cells (Eguchi 1998).

The inhibitory influence of the injury environment on regeneration-competent cells has been shown not only in the regenerating spinal cord (Filbin 2000), but in experiments testing the effects of skin wound fluids on rat myoblast proliferation and differentiation (Sicard and Nguyen 1994). Early wound fluids obtained from subcutaneous polyvinyl alcohol (PVA) sponges 1 and 3 days after implantation stimulated myoblast proliferation in vitro, but later wound fluids obtained 5, 10, and 15 days post-implantation showed decreasing ability to support myoblast proliferation and inhibited myoblast differentiation. Myoblast fusion was decreased by 88%–100% and creatine kinase activity was depressed by 60%–75%. Myoblasts implanted alone into subcutaneous PVA sponges survived in the presence of granulation tissue, but did not fuse into myotubes, whereas they did fuse in the presence of minced muscle regenerating within the sponges (Sicard et al. 1997). Granulation tissue formed in PVA sponges loaded with minced muscle exhibited increased cellularity and reduced collagen deposition.

2.2.1.3
Induction of Dedifferentiation of Mouse Myotubes

Mouse myofibers do not cellularize and dedifferentiate after injury in vivo, but they have the latent ability to do so, as shown by two experimental manipulations of

C2C12 myotubes in vitro. First, the gene *msx1*, which is expressed in the undifferentiated limb bud mesenchyme and in regenerating amphibian limbs (Hill 1989; Robert et al. 1989; Simon et al. 1995; Crews et al. 1995), inhibits myogenesis in vitro when expressed constitutively in cultured mouse myoblasts (Song et al. 1992; Woloshin et al. 1995). Forced expression of *msx1* in cultured C2C12 myotubes induces their morphologic and molecular dedifferentiation, reducing the expression of muscle regulatory proteins to undetectable levels in 20%–50% of the myotubes (Odelberg et al. 2001). About 9% of the dedifferentiated myotubes cleave to produce either smaller multinucleated myotubes or proliferating mononucleate cells. Clonal populations of these mononucleate cells are multipotent, able to differentiate in vitro into cells that express chondrogenic, adipogenic, myogenic, and osteogenic molecular markers, like mesenchymal stem cells of the bone marrow stroma (Pittinger et al. 1999; Odelberg et al. 2001).

Second, morphological dedifferentiation of C2C12 myotubes is induced by a protein extract derived from amputated newt limbs undergoing dedifferentiation (McGann et al. 2001). Muscle differentiation proteins were reduced to undetectable levels in 15%–30% of the treated myotubes, and 18% exhibited entry of nuclei into S phase. Of the myotubes, 11% cleaved and about 50% of these continued cleaving to produce proliferating mononucleated cells. Control myotubes exposed to extract of unamputated limbs did not cleave and dedifferentiate. These data indicate that mouse myotubes retain much of the mechanism of muscle dedifferentiation observed in regenerating newt limbs (Brockes and Kumar 2002) and that the newt regeneration extract supplies the mouse myotubes with the missing part of the mechanism.

2.2.2
Interventions to Promote Regeneration In Vivo

A number of attempts to improve or induce regeneration of various tissues have been made, with variable degrees of success.

2.2.2.1
Peripheral Nerve

When a peripheral nerve is transected, that portion of the axons distal to the cut die, and the proximal axon stumps sprout and regenerate through the now-empty distal endoneurial tubes (Yannas 2001). However, if a gap is created in a nerve, neighboring fibroblasts invade the gap and form scar tissue that prevents distal regeneration, with resultant loss of sensory and motor function. Regeneration of the nerve stump can be promoted by bridging the gap with a nerve autograft, muscle, or blood vessels, but this technique means loss of tissue at the donor site.

Tubular nerve guides made of silicone or ethylene-vinyl acetate (EVAc) have been designed to promote the regeneration of rat sciatic nerve across a gap (Bellamkonda and Aebischer 1995). The ends of the severed nerve are inserted into the guide, which directs the regenerating axons toward their distal targets and prevents scar tissue from invading the area of regeneration. The texture of the guide wall, its porosity and the presence of ECM molecules, growth factors, or Schwann cells all have been shown to influence the regeneration process (Yannas 2001). Smooth guide tube walls promote straight growth of neurites better than rough ones. The porosity of the walls has to be large enough to allow exchange between the inside of the guide and the external environment. Biodegradability of nerve guide tubes is a desirable feature; otherwise the guides must be removed from the regenerated nerve by a second surgical operation. Biodegradable guides need to survive for ~4–12 weeks and their degradation products should not have an adverse effect on regeneration.

Silicon filled with a collagen/chondroitin 6-sulphate matrix or an agarose matrix containing the laminin recognition peptide CDPGYIGSR or FGF promote sciatic nerve regeneration across large gaps (Aebischer et al. 1989; Bellamkonda and Aebischer 1995; Yannas 1995). Similarly, seeding silicone guide tubes with Schwann cells enhances regeneration (Guenard et al. 1992). Porous collagen, poly-L-lactic acid and polyglycolic nerve guides promote regeneration of the sciatic nerve (Henry et al. 1985; Molander et al. 1989; Yannas 1995, 2001). The ultimate goal would be to design a nerve guide made of a biodegradable polymer with covalently linked matrix recognition sequences, and impregnated with neurotrophic factors tailored to the motor (NTF-4/5, CNTF, GDNF, or FGF-5) or sensory (NGF, BDNF) quality of the nerve. The neurotrophic factors would be slow-released as the guide degrades.

2.2.2.2
Spinal Cord Injury

Interventions that have been used to prevent paralysis and promote regeneration in injured spinal cord are neuroprotective and survival agents, neutralization of proteins that inhibit axon extension, inhibition of gliosis, enzymatic digestion of glial scar ECM, and implanting scaffolds that promote axon extension.

The only currently FDA-approved neuroprotective agent for human spinal cord injury is methylprednisolone. This steroid stabilizes cell membranes, and its use doubles the chances for functional recovery if administered within the first 8 h after injury (Bracken et al. 1990, 1992). Gacyclidine, a non-competitive antagonist of the glutamate receptor N-methyl-D-aspartate, reduces glutamate release and improves recovery from spinal cord contusion injury in rats (Feldblum et al. 2000; Gaviria et al. 2000). Polyethylene glycol (PEG), in conjunction with Fampridine (4-aminopyridine), induces the recovery of action potentials in guinea pig spinal cord after compression injury to 72% of pre-injury levels (Luo et al. 2002). Other agents that have been shown to have neuroprotective effects in animal models are dexamethasone, N-methyl-D-aspartate, α-amino-3-hydroxy-5-methyl-4-isoazolepropionate kainate (Gonzalez et al. 1996; Kaku et al. 1993) and water-soluble carboxy "buckyballs," spheres of carbon to which are attached six pairs of carboxylic acid molecules. Buckyballs are proficient at soaking up toxic free radicals generated by injury. When tested on oxygen and glucose-starved neurons in vitro (conditions that mimic the effects of stroke), neuron death was reduced by 75% (Dugan et al. 1997).

Growth and neurotrophic factors that promote the survival of peripheral neurons during ontogenesis and regeneration also enhance the survival of injured central neurons. Schnell et al. (1994) have shown that injection of NGF or neurotrophin-3

(NT-3), but not brain-derived neurotrophin (BDNF), increases the initial sprouting of injured axons, but the distance of growth is small. Transduction of spinal motoneurons with the NT-3 gene in an adenoviral vector resulted in a significant increase in the concentration of NT-3 in the L3–L6 region. This increase was associated with the ability of axons to grow from the intact corticospinal tract after hemisection of the opposite tract (Zhou et al. 2003). FGF-2 may also prove useful as a neural survival factor. Four days after contusive spinal cord injury in rats there is a significant increase in FGF-2 expression by glial cells that may aid in functional recovery (Mochetti et al. 1996).

Glial scar formation is an important physical and chemical barrier to axon regeneration. Wamil et al. (1998) reported that application of CM101, a polysaccharide derived from group B streptococcus, to the lesioned spinal cords of rats prevents angiogenesis and the infiltration of inflammatory immune cells, thus limiting glial scar formation. The survival rate of the animals was high and they regained the capacity to walk. Treatment of lesions with chondroitinase ABC resulted in the recovery of walking pattern and the ability to walk a beam without foot slipping in rats that were paralyzed by crushing the dorsal columns at the level of the fourth vertebra (Bradbury et al. 2002). This enzyme removes chondroitin sulfate glycosaminoglycans (GAGs) from the core protein of CS proteoglycans. These GAGs are anti-adhesive and would normally inhibit axon extension across the scar, but their removal renders them adhesive. The large degree of functional motor recovery indicates that axons regenerating from neurons in the motor cortex made a significant number of functional synapses on the other side of the lesion.

Chondroitin sulfate proteoglycans (Niederost et al. 1999) and myelin proteins in spinal cord lesions (Filbin 2000) inhibit the sprouting of injured axons and constitute part of the glial scar. Two myelin proteins that have been identified as inhibitory are Nogo (Caroni and Schwab 1988; GrandPre et al. 2000; Prinjha et al. 2000) and myelin-associated glycoprotein (MAG) (Filbin 2000). Nogo and MAG are expressed by oligodendrocytes and collapse the growth cones of axons by binding to their receptors on regenerating axons. Treatment of lesioned rat spinal cord with monoclonal antibodies to Nogo resulted in axon regeneration; however, the number of axons that regenerated was low and few or no functional synapses were made (Schnell and Schwab 1990). Combining anti-Nogo antibody treatment with injection of NT-3 enhanced corticospinal axon regeneration, so that 5%–10% of the corticospinal fibers were regenerated in successful cases (Schnell et al. 1994). The cord was only partially transected in these experiments, however, and the regenerated axons went through the intact part of the cord, not through the lesion itself. Another approach is to block the action of the Nogo and MAG receptors. Synthetic amino-terminal peptide fragments of the 66-residue domain of Nogo are competitive antagonists of Nogo for the Nogo receptor and promoted axon regeneration in rats when administered to lesions created by severing the dorsal halves of the corticospinal tracts (GrandPre et al. 2002). Growth factors such as BDNF, GDNF, and NGF block the inhibitory effect of MAG by blocking the intracellular inhibitory pathway triggered by MAG when it binds to its receptor on axons (Cai et al. 1999).

There are likely to be at least several other inhibitory myelin proteins to be identified, as shown by the fact that vaccination of mice with preparations of spinal cord enriched in myelin results in better regeneration after dorsal hemisection of corticospinal tracts than treatment with antibody to Nogo, (Huang et al. 1999). Retrograde and anterograde labeling indicated that numerous axons regenerated across the lesion, many through the glial scar. At least some of these axons appeared to make functional synapses, since 58% of the vaccinated animals showed functional recovery, defined as the ability of the animals to lift their foot and place it on a support surface when the dorsal surface of the hindlimb was touched. Coordinated locomotion, however, was not reported. Still to be resolved is the question of whether the antibodies resulting from vaccination are required just for the initial stages of axon extension, or need to be present for the whole course of regeneration.

Functional axonal regeneration of spinal cord axons takes place in neonatal rats after transplanting fetal cord to bridge the lesion, as demonstrated by anterograde and retrograde labeling of axons with fluorescent dyes, and by the recovery of righting reflexes and ability to walk (Iwashita et al. 1994). Fetal spinal cord implants do not induce the functional regeneration of adult spinal cord, but adult spinal cord axons have been induced to regenerate across lesions by implanting segments of intercostal nerves embedded in a fibrin matrix impregnated with FGF-1 (Cheng et al. 1996). The endoneurial tubes of the nerve segments bridged white and gray matter, allowing host axons to regenerate through them without being impeded by inhibitory proteins and glial scar. Anterograde and retrograde labeling showed that the axons regenerated from the white matter through the endoneurial tubes to the gray matter. There was some functional recovery; the rats were able to support their weight on the paralyzed hindlimbs and displayed movement in all leg joints, but did not recover co-ordinated locomotion. The inclusion of FGF-1 in the fibrin matrix was essential for regeneration. No improvement was noted when the matrix lacked FGF-1, indicating that this growth factor may be essential for axon extension.

As might be expected, olfactory ensheathing cells (OECs) also support regeneration of spinal cord axons in rats (Ramon-Cueto et al. 1998). The spinal cord was transected at the thoracic level and bridged with a PVC tube containing Schwann cells imbedded in Matrigel. The Schwann cells were included to provide guidance factors for regenerating axons. Olfactory ensheathing cells were injected on either side of the bridge, where they migrated into the cord stumps and the Matrigel. There was minimal axon regeneration into the Matrigel in the absence of OECs, but long-range regeneration in both directions across the Matrigel was observed. Interestingly, glial scar formation was not inhibited, but the regenerating axons penetrated the glial scar. The OECs might function in one or both of two ways to promote regeneration. First, they are known to express adhesion molecules, particularly laminin, that are important for axon guidance and extension, as well as growth factors such as PDGF and neuropeptide Y that are essential for neuron survival. Second, they might interact with astrocytes and oligodendrocytes in the glial scar to alter the molecular expression of these cells, thus modifying the molecular composition of the scar from inhibitory to more permissive. No functional recovery tests were reported in this experiment.

Goldsmith and de la Torre (1992) implanted a collagen matrix containing laminin or 4-aminopyridine into cat spinal cords (six animals in each group) following complete transection (Goldsmith and de la Torre 1992). Regeneration of supraspinal fibers was reported for a distance of up to 90 mm below the transection site. The synaptogenic marker, synaptophysin, was detected on preganglionic sympathetic neurons in association with dopaminergic and noradrenergic-containing varicosities below the collagen bridge, suggesting that synaptic connections had been made with these sympathetic neurons. One cat in each group showed recovery of coordinated locomotion, the remainder did not.

Other CNS regeneration templates that have been tried are collagen containing genetically modified Schwann cells to hypersecrete NGF (Weidner et al. 1999), alginate gel (Kataoka et al. 2000; Suzuki et al. 2000), alginate gel containing fibroblasts genetically modified to secrete BDNF (Tobias 2000), and "Neurogel," a hydrogel incorporating the RGD adhesive sequence (Woerly et al. 2000). In most of these studies, functional recovery was not measured or the recovery was small.

Neonatal rats can regain the ability to walk without intervention after complete spinal cord transection. Recovered rats regenerate those nerve tracts that control posture, balance, and coordination, but do not regenerate the corticospinal tract, which is the primary conduit for signals controlling voluntary movements (Wakabayashi et al. 2001). This means that there is a tremendous amount of synaptic plasticity in the tracts that control non-voluntary motor functions, allowing them to compensate for the inability of the corticospinal tract to regenerate. Synaptic plasticity involves creating and eliminating synapses among the neurons still present after an injury, as well as strengthening or weakening of existing synapses. There is substantial evidence that this plasticity takes place where dendritic spines make excitatory synapses with axons and involves changes in spine morphology based on the dynamic reorganization of actin filaments in the spines and influenced by trophic factors, hormones, and astrocyte activity (Matus 2000). Electron microscopic studies of dendritic segments of pyramidal neurons in the mouse barrel cortex showed that, while dendritic structure is stable, dendritic spines appear and disappear, depending on changing sensory experience (Trachtenberg et al. 2002). Parallel to morphallaxis, I call this changing synaptic reorganization "synaptallaxis".

Another factor that can induce synaptic plasticity is muscle use, through either exercise regimens or functional electrical stimulation. Recovery from spinal cord injury is possible in younger animals by using these means, particularly in cases of incomplete injury (Peckham and Creasy 1992), but becomes poorer with age. For example, Muir (1999) showed that chicks learned to walk on their own after hemisectional thoracic injury, but could not relearn swimming movements. By using stimulation of the feet that made their legs move in a swimming pattern, the chicks were able to make leg movements that approximated the swimming motions made prior to hemisection. In humans with spinal cord injury, treadmill exercises involving repetitive coordinated motion have been of some help in re-training the spinal cord for walking motions, but is not completely successful. Electrical stimulation of muscles in patterns that mimic coordinated movements also helps regain some motor function and improve respiratory, bladder, bowel, and sexual function (Peckham and Creasy 1992).

Few of the interventions used for spinal cord injury in experimental animals have resulted in major, or even consistent, recovery from paralysis. Protocols for experiments are highly variable. Tests for functional recovery often are not part of studies or the tests used vary widely from one study to another. Furthermore, there are differences between mice and rats, as well as differences among strains of animals, in their response to neural trauma (Steward et al. 1999). Thus, many studies have not been comparable or reproducible. Nevertheless, a major accomplishment has been the induction of axon regeneration across lesions. What is still missing is the growth of substantial numbers of axons into healthy tissue on either side of the lesion, the establishment of new synapses and reorganization of old ones that results in the restoration of a functional circuitry that leads to recovery. Hopefully, further research combining a number of the approaches that have been successful in promoting axon regeneration and synapse formation and reorganization will result in success (McDonald and Sadowsky 2002).

2.2.2.3
Parkinson's Disease

Genetic modification and growth factor activation of resident stem cells have been tried as therapies for Parkinson's disease. Kordower et al. (2000) induced symptoms of the disease in monkeys by intracarotid injections of 1-methyl-4-phenyl-1,2,3,6-tetrahydropyridine (MPTP), followed by injections into the substantia nigra and striatum of lentiviral constructs containing the gene for glial-derived neurotrophic factor (GDNF). Three months later, motor performance of the monkeys was vastly improved, and PET scans with ^{18}fluorodopa showed an uptake of the tracer over 300% greater than in control animals injected with constructs containing the β-galactosidase gene. Histological studies revealed many GDNF$^+$ neurons in the substantia nigra and striatum and TH$^+$ neurons were abundant. The mechanism of the GDNF effect is unknown, but is likely to be neuroprotection.

Cells of the subventricular zone proliferate in vitro in response to TGF-α, suggesting that this growth factor might be used to activate NSCs in the walls of the lateral ventricles of unilateral Parkinsonian rats that would then migrate to the degenerating substantia nigra and differentiate into DANs. A total of 50 μg of TGF-α was administered to the striatum via a shoulder-implanted minipump and cannula at the rate of 0.5 μl/h for 2 weeks (Fallon et al. 2000). The infused rats showed a 31.5% improvement in rotational behavior over controls. Studies of BrdU incorporation and immunostaining for nestin and differentiated neuron markers revealed that NSCs proliferated, migrated to the striatum and differentiated into new neurons, some of which were presumably DANs.

2.2.2.4
Bone and Cartilage

Bone cannot regenerate across a large, surgically created gap, and fractures in which the ends of the broken bone are too far displaced from one another result in the formation of scar tissue instead of reunion by bone regeneration. There would appear to be a minimum distance between the edges of a bone defect beyond which regeneration will not occur. The failure of regeneration in large bone defects might be because the synthesis of essential ECM molecules and release of growth factors from bone matrix, macrophages, and osteoclasts is localized to the ends of the bone, and thus fails to induce bone formation beyond the edges of the defect. Consistent with this idea, matrices impregnated with BMPs have been found to induce regeneration from host bone when implanted into surgically inflicted defects in rat and sheep femur, and dog mandible, that normally are too large to elicit regeneration (Rosen and

Theis 1992). Ceramic matrices can induce bone regeneration across a large gap without the inclusion of growth factors. One such matrix is a paste consisting of monocalcium phosphate monohydrate, α-tricalcium phosphate, and calcium carbonate, to which is added a solution of sodium phosphate (Constanz et al. 1995). When injected into gaps in a bone, the paste rapidly hardens into a material resembling bone matrix that is invaded by osteoblasts from the host bone, which remodel it into natural bone matrix. Another matrix that induces bone regeneration across gaps is a composite of poly (propylene-fumarate), crosslinked through the fumarate double bond by a monomer, N-vinylpyrrolidone, and filled with NaCl and phosphate β-tricalcium (Yaszemski et al. 1995). This composite is plastic when mixed fresh and can be pressed into bone defects. When hardened, it is strong enough to substitute temporarily for bone, and it degrades as osteoblasts invade it and regenerate new bone. In both these cases, it is likely that autocrine growth factors produced by the osteoblasts are playing a major role in their proliferation and differentiation as they invade the artificial matrix.

Attempts to induce the regeneration of articular cartilage have been largely unsuccessful. The use of a growth factor-impregnated fibrin matrix, in combination with protease treatment failed to stimulate regeneration in surgically inflicted, partial-thickness defects in the articular cartilage of rabbits and minipigs (Hunziker and Rosenberg 1996). The wound surface was treated for 5 min with 1 unit/ml of chondroitinase ABC to expose the collagen network of the ECM, providing an adhesive surface for the migration of mesenchyme-like cells from the synovium into the wound. Protease treatment alone resulted in the migration of a few mesenchymal cells into the defect. Migration was further enhanced by the application of TGF-β1 or FGF-2 to the wound. Maximum migration was achieved by filling the protease-treated wound with a fibrin clot impregnated with TGF-β1 or FGF-2. The fibrin clot degraded, and by 5 weeks in rabbits and 7 weeks in minipigs, was completely replaced by a loose connective tissue with a matrix consisting predominantly of collagen fibrils. No new cartilage was formed, however, even after 48 weeks.

2.2.2.5
Skin Wounds

Biodegradable, cell-free artificial regeneration templates induce dermal regeneration in excisional skin wounds, though the regeneration is imperfect. These templates typically consist of collagen or gelatin sponges to which other ECM components, such as chondroitin 6-sulfate or elastin, are added (Yannas 2001). The matrix is overlaid with either a meshed split-thickness epidermal graft or a suspension of dissociated keratinocytes. Fibroblasts from the surrounding dermis migrate into the artificial matrix and replace it with a matrix that resembles, but is not identical to, normal dermal matrix. One of the best results was obtained with a dermal scaffold made of type I collagen and elastin applied to a full-thickness excisional wound in pig skin (DeVries et al. 1994). The regenerated skin was smooth and had the same color as the surrounding skin. Its histological appearance approximated that of normal skin with collagen fibrils oriented in the typical basket-weave pattern. Hair follicles, sweat glands, and sebaceous glands did not form.

It is not known whether the responding cells in dermal regeneration experiments are the same fibroblasts that normally would form scar tissue or are a subpopulation of regeneration-competent reserve cells that are suppressed in favor of repair. There is some evidence, based on differences in thymidine incorporation and pattern of cell association in vitro, that hypodermal fibroblasts form scar, whereas dermal fibroblasts might be capable of regeneration (Gross 1996). The presence of an artificial dermal matrix in a wound might suppress wound contraction and proliferation of hypodermal fibroblasts and promote proliferation of dermal fibroblasts, leading to regeneration. The factors that suppress dermal fibroblast proliferation after wounding or promote hypodermal fibroblast proliferation in the presence of an artificial dermal matrix are unknown, but could be investigated in vitro with an artificial matrix system.

2.2.2.6
Use of Pig Small Intestine and Urinary Bladder Submucosa as a Regeneration Template

Two interesting cell-free regeneration templates for the promotion of regeneration are pig small intestine submucosa (SIS) and pig urinary bladder submucosa (UBS) (Badylak 2002). To prepare these templates, the serosa and external muscularis mucosa, and the tunica mucosa are removed, leaving the submucosa, muscularis mucosa and attached stratum compactum. The remaining tissue is processed to eliminate cells. The SIS template, which has been analyzed most extensively, is composed mainly of collagen and fibronectin, but also contains hyaluronic acid, heparin, heparan sulfate, chondroitin sulfates A and B, and dermatan sulfate (Hodde et al. 1996, 2002). It does not elicit a strong immune response when implanted into a host and is more resistant to infection than Teflon grafts (Badylak et al. 1994).

SIS or UBS is effective in promoting the regeneration of skin (Prevel et al. 1995), esophagus (Badylak et al. 2000), dura mater (Cobb et al. 1999), blood vessels (Badylak et al. 1989; Sandusky et al. 1995; Prevel et al. 1994), urinary bladder (Merguerian et al. 2000; Probst et al. 2000; Reddy et al. 2000), tendon (Aiken et al. 1994; Hodde et al. 1997) and ligament (Lantz et al. 1993; Hiles et al. 1995; Badylak 2002). The templates are invaded by host cells that become progressively organized into new tissue. The new tissue produces its own extracellular matrix while the template degrades. SIS is now in widespread clinical use to repair these tissues.

2.2.3
Strategies to Define the Molecular Requirements for Chemical Induction of Regeneration

There is every reason to believe that if we know the correct combination of signals and their receptors, we should be able to not only activate stem cells that are resident in tissues, but to create stem cells by the reprogramming of differentiated cells that leads to their dedifferentiation. How might we acquire this combination? A direct strategy is to compare and contrast the patterns of gene activity in regeneration-competent vs regeneration-deficient tissues to define the molecular signals, receptors, and injury products that distinguish regeneration-permissive from regeneration-inhibiting conditions. Three types of comparative models are useful: (1) tissues of mutant vs normal animals of the same species; (2) tissues of regeneration-competent vs deficient stages of the life-cycle in the same species; and (3) tissues of regenerating vs non-regenerating species. The data obtained can then be used to design molecular "cocktails" of genes or proteins to mimic an injury environment permissive for regeneration and/or render cells competent to respond to a permissive injury environment.

A comparative strategy has been used to define a common set of genes for "stemness" in ESCs and ASCs (Ivanova et al. 2002; Ramalho-Santos et al. 2002). Transcriptional profiles of mouse ESCs, NSCs, HSCs, and differentiated cells from the lateral ventricles of the brain and the main population of bone marrow were compared by hybridizing DNA microarrays containing several thousand genes with mRNAs from each set of cells. Bioinformatic analysis was used to identify transcripts absent in differentiated cells but present in stem cells and to assign transcripts enriched in stem cells to functional categories. Both studies found a common set of 216 (Ramalho-Santos et al. 2002) and 283 (Ivanova et al. 2002) genes enriched in all three sets of stem cells that appear to define core "stemness." These genes encode proteins for (1) various aspects of signaling (JAK/STAT transducers and activators of transcription and Notch signaling, ability to sense growth hormone and thrombin, and interaction of cell adhesion molecules with ECM); (2) entrance and progression through the cell cycle; (3) high resistance to stress, with upregulation of DNA repair, protein folding, ubiquitin system, and detoxifier systems; (4) transcriptional regulation, including chromatin remodeling; and (5) translational regulation. More than 50% of the commonly enriched genes were ESTs, which represent unidentified genes. Otherwise, the gene activity of each set of ESCs is distinct from one another. HSCs are more similar to the main bone marrow population than to ESCs or NSCs, but NSCs are more similar to ESCs than to HSCs or differentiated cells (Ramalho-Santos et al. 2002).

2.2.3.1
Comparison of Mutant Vs Wild-Type Tissues

The MRL/lpr mouse displays gross lymphoproliferation and autoimmune disease, and has been used as an experimental model for study of the immune system. Immunologists identify mice in experiments by ear punch-holes. In most mice, such as C57B1/6, the punch holes heal like standard excisional skin wounds: they re-epithelialize, form scar tissue at the rim, and remain open. However, punch holes in the ears of the MRL/lpr mouse close completely in less than 4 weeks. Histological studies show that closure is accomplished by a blastema of proliferating cells at the wound edges. The blastema regenerates the normal structure of the ear tissue, including the supporting cartilage sheet (Desquenne-Clark et al. 1998). A similar result is observed with punch holes in the ears of MRL/Mp mice (Masinde et al. 2001).

The capacity for mouse ear tissue regeneration is quantitative and associated with a heritable multigenic trait. To identify the genetic loci underlying the regenerative ability of these mutant mice, a genome-wide scan using microsatellite DNA markers was performed on the F2 intercross progeny of regenerating and non-regenerating mouse strains, MRL/lpr×C57Bl/6 J (McBrearty et al. 1998) and MRL/MP×SJL/J (Masinde et al. 2001). A normal quantitative distribution of regeneration was exhibited by the F2. Thirteen quantitative trait loci (QTL) associated with a high degree of regeneration have been identified on chromosomes 1, 4, 6, 7, 8, 9, 12, 13, and 15. Eleven of these were inherited from the regenerating MRL mice and two from the non-regenerating strains. These loci explain 70% of the variance in regeneration of the F2 mice (Masinde et al. 2001). Several of the QTLs appear to interact during regeneration (McBrearty et al. 1998; Masinde et al. 2001). Several genes known to be involved in limb development and regeneration have been identified at the QTLs: FGFR4; Gli3, the transcription factor activated by sonic hedgehog; the *Hoxc* cluster; and RARγ, the mammalian equivalent of the newt RARδ, the receptor that mediates retinoic acid-induced changes in blastema cell positional identity in regenerating urodele limbs (Pecorino et al. 1996). Many other genes at these loci are involved in signal transduction pathways (McBrearty et al. 1998).

Regeneration of cardiac muscle has also been investigated in the MRL/MpJ mouse (Leferovich et al. 2001, 2002). The myocardium of the left ventricle was cryo-injured transmurally and the course of repair followed for 60 days. C57/BL/6 hearts served as controls. DNA synthesis and mitosis were assessed by BrdU incorporation and DAPI staining. The infarct filled with scar tissue in the control mice. In the mutant mice, confocal microscopy demonstrated cells in which antibody staining for α-actinin and BrdU were colocalized, indicating that cardiomyocytes were proliferating and migrating into the infarct to regenerate cardiac muscle. These cells had a mitotic index of 10%–20%, compared to 1%–3% in controls. Revascularization was more evident at the injury site in the MRL mice. By day 60, the mutant hearts were virtually scar-free and echocardiographic measurements indicated a return of heart function to normal.

2.2.3.2
Comparison of Regeneration-Competent Vs Regeneration-Deficient Stages of the Life Cycle

Embryonic tissues are known for their ability to regulate after removal of cells or scrambling of cells within the rudiment. This regulative ability extends into the fetal stages of mammalian development in the form of regeneration of the skin (Whitby and Ferguson 1991; Mast et al. 1992; Olutoye and Cohen 1996; McCallion and Ferguson 1996), into tadpole stages of frog limb development as the ability to regenerate spinal cord, limb and tail, and into adult stages of urodeles as the ability to regenerate numerous complex structures.

Fetal Vs Adult Mammalian Skin. Fetal mammalian skin is regenerated rapidly and perfectly after excisional wounding. Mouse embryo (11.5 days of gestation) cultures have been used to study the closure of excisional wounds created by amputating the hindlimbs and growing the embryos in roller tubes for 48 h (McCluskey and Martin 1995). Epithelialization of the amputation surface is complete by 24 h. Confocal microscopy revealed that re-epithelialization occurs not by lamellipodal crawling (migration) as in the adult, but by a purse-string contraction of a filamentous actin cable assembled in the basal layer of cells at the edges of the wound that is blocked by cytochalasin D. Labeling of mesenchymal cells at the wound edge with DiI showed that the dermis of the wound moves inward by 50% within 24 h. This contraction is not due to the conversion of mesenchymal cells to myofibroblasts as in adults, but is due to the active movement of mesenchymal cells over the ECM substrate (Mast et al. 1997). Similar results were obtained on wounded fetal rat skin (Ihara and Motobayashi 1992). No granulation tissue is formed in fetal wounds. Late in gestation, the skin changes its injury response from regeneration to fibrosis. In the rat and mouse, this transition takes place at 16–18 days of gestation. Wounding initiates an inflammatory response that leads to production of granulation tissue that is subsequently remodeled into scar by lysyl oxidase-catalyzed crosslinking of collagen I fibrils.

Protein synthesis is elevated above normal in both fetal and adult wounds. However, there is a greater collagen/total protein ratio in the adult wound vs normal skin, whereas there is no change in this ratio between unwounded and wounded fetal skin (Houghton et al. 1996). The same collagens are synthesized by fibroblasts in fetal and adult wound, but there is no excessive deposition of type I collagen fibrils in fetal wounds and the fibrils exhibit the normal reticular pattern (Frantz et al. 1992; Mast et al. 1997). The normal pattern of collagen deposition and organization in fetal wounds is correlated with three fetal-specific ECM features which diminish fibroblast proliferation and crosslinking of type I collagen fibrils. First, fetal wound fibroblasts synthesize higher levels of HA and HA receptor than adult wound fibroblasts (DePalma et al. 1989; Longaker et al. 1993; Alaish et al. 1994), giving them more opportunity to bind HA. HA has been shown to inhibit fetal fibroblast proliferation (Mast et al. 1993) and to decrease scar formation in wounds of adult tympanic membranes (Hellstrom and Laurent 1987). Conversely, treatment of fetal rabbit skin wounds with hyaluronidase or HA degradation products alters the regenerative response toward fibrosis (Mast et al. 1992, 1995). Second, sulfated proteoglycan synthesis does not accompany collagen synthesis in fetal wounds (Whitby and Ferguson 1992). Third, fetal skin has a higher ratio of type III to type I collagen (Epstein 1974; Merkel et al. 1988).

Differences in the organization of fetal and adult wound ECM may in turn be due to differences in the inflammatory response. Both TGF-β and FGF-2, which stimulate the formation of granulation tissue in adult wounds, are absent from unwounded and wounded fetal rat skin, although PDGF is present and fetal fibroblasts have functional TGF-β receptors (Whitby and Ferguson 1991; Sullivan et al. 1993). The absence of these growth factors is correlated with the virtual absence of an inflammatory response compared to that of adults (Cowin et al. 1998). Only small numbers of monocyte-derived cells are recruited into fetal mouse skin wounds by 3 h postwounding, and no activated macrophages appear until 18 h. All types of macrophages are eventually recruited to the fetal wound but their numbers and persistence are much lower than in adult wounds. Activated macrophages release signals that recruit B cells in adult wounds. B cells express IL-6, which works synergistically with IL-1 to recruit fibroblasts, thus helping to maintain an environment that promotes production and deposition of fibrotic matrix. Because the macrophage presence in fetal wounds is so diminished, B cells do not appear in fetal wounds. However, placing pellets of bacteria in PVA sponges subcutaneously in fetal rabbits elicits an acute inflammatory response that initiates an adult-like fibrosis. This response includes recruitment of neutrophils and large numbers of leukocytes (Frantz et al. 1993).

A lack of macrophages would also reduce TGF-β levels in fetal wounds. Consistent with this idea, reducing the levels of TGF-β1 and 2, but not PDGF or FGF, in vivo or in vitro by application of neutralizing antibodies reduces scarring in adult wounds (Shah et al. 1992, 1994; Houghton et al. 1995). Conversely, treating fetal rodent wounds with exogenous TGF-β induces scarring (Krummel et al. 1988). Reduction in scarring is enhanced by applying the antibody immediately after injury, suggesting that the effect of lowering TGF-β levels at the start of the injury response cascades all the way through the repair process. At high concentrations, PDGF also causes a shift to a more adult-like pattern of repair in fetal skin wounds (Haynes et al. 1990).

The shift from a regenerative to a scarring response may reflect intrinsic changes in fibroblasts in addition to environmental changes. In cultured 14-day mouse limbs, the transition from regeneration to scarring in incisional wounds takes place autonomously, in the absence of immune cells and systemic factors (Chopra et al. 1997). Fetal fibroblasts maintain their regenerative response when grafted subcutaneously into adult athymic mice (Lorenz et al. 1992; Lin et al. 1994) and they have a unique phenotype differing from that of the adult fibroblast, including production of and response to growth factors, synthesis of matrix molecules, pericellular HA coats and antigen determinants (Moriarty et al. 1996; Ellis et al. 1997; Gosiewska et al. 2001). Another possibility is that the transition to adult fibrosis might involve the loss of a specific fibroblast cell type capable of a regenerative response.

It is important to note that regeneration of fetal mammalian wounds is tissue-specific; not all fetal tissues are able to regenerate. For example, in fetal sheep the skin regenerates, but injured diaphragm muscle exhibits a combination of regeneration and fibrosis, and gastrointestinal tract wounds heal by fibrosis (Longaker et al. 1991; Meuli et al. 1995). Furthermore, there appear to be species differences in the way fetal wound healing is regulated in mammals. Longaker et al. (1994) reported that macrophages are recruited into fetal sheep skin wounds and that TGF-β2 is present in higher concentration in the fetus than in the adult, just the reverse of the findings in rodents. Adult mammalian saliva contains high levels of TGF-β2, yet oral lesions in adults heal with little or no scarring, suggesting that inhibitors of TGF-β such as decorin and α2-macroglobulin could play a role in minimizing scarring (Danielpour et al. 1990; Yamaguchi et al. 1990).

Fetal Vs Adult Mammalian Cartilage. The ear cartilage of wild-type adult rats does not regenerate. After making an incisional wound all the way through the ear, a fibrotic scar forms between the edges of the incised cartilage that stains intensely with antibody to type I collagen. Incisions in the cartilage of neonatal rat ears, however, are repaired by the regeneration of chondrocytes that express type II collagen. Nuclear PCNA staining indicated that the regenerated chondrocytes are derived by the proliferation of neonatal chondrocytes in the lesion area (Wagner et al. 2001).

The articular cartilage of joints is typical hyaline cartilage with a matrix consisting of hyaluronic acid, proteoglycans, and type II collagen (Ham and Cormack 1978). Roughly 80% of the wet weight of cartilage is water, due to the high water binding capacity of HA in the matrix. The structure of hyaline cartilage matrix gives it its physical properties of elasticity, firmness, and resistance to deformity, properties essential for its weight bearing function. Articular cartilage is avascular, but because of the high water content of its matrix, it is easily nourished by diffusion of oxygen and nutrients from the synovial fluid of the joint capsule. Mitosis of superficial chondrocytes ceases in the adult, and thus the only way that adult articular cartilage can compensate for wear is by the production of new matrix.

Two major retrogressive changes occur in articular cartilage with age (Stockwell 1979). The first is calcification, which decreases the diffusion of nutrients and oxygen to chondrocytes. Chondrocytes of the calcified matrix die and the matrix is resorbed. The second change is termed cartilage fibrillation and leads to osteoarthritis. Fibrillation involves splitting of the articular surface along the direction of orientation of the collagen fibers. The fibers are exposed, giving the articular surface a fuzzy appearance in sections for microscopy. Fibrillation occurs at first in patches, which then enlarge. As the condition progresses, there can be variable loss of cartilage and exposure of underlying bone, and pain. The cause of fibrillation is unknown in most cases.

The reparative ability of adult articular cartilage is low (Campbell 1969). Damage to the cartilage that does not involve the underlying bone does not heal, or the lesion is filled with fibrocellular (scar) tissue. If the articular cartilage and underlying bone are injured simultaneously, however, the defect in the cartilage will be imperfectly repaired by MSCs as the bone is repaired. In the avascular environment of the cartilage, the MSCs differentiate into fibrocartilage cells, whereas in the vascular environment of the bone, they differentiate into osteoblasts. Cartilage repair is of higher quality if the joint is kept in motion during the process, probably because the motion of the synovial fluid provides for better nutrient supply and waste removal than if the joint is immobilized (Salter et al. 1978). Rheumatoid arthritis is an autoimmune disease that also destroys articular cartilage. Autoantibodies attack an as-yet-unknown synovial antigen, which leads to a chronic inflammatory response that destroys the cartilage (Janeway 1997).

Namba et al. (1998) have shown that partial thickness incisions in the articular cartilage of fetal lambs are healed by the proliferation of chondrocytes in the lesion area, much like the regeneration of cartilage in neonatal rat ears (Wagner et al. 2001). Clearly, fetal articular chondrocytes and neonatal ear chondrocytes are still able to respond to injury by regeneration. The molecular environments of fetal/neonatal vs adult cartilage have not been as fully explored as in fetal and adult skin.

Early Vs Late Stage Frog Limb Buds. Anuran limb buds regenerate well, but lose this ability as the limb differentiates and the animals approach metamorphosis. This loss of regenerative capacity has been best studied in the hindlimbs of the frog *Xenopus laevis*. The hindlimb buds of *Xenopus* begin to differentiate proximally at stage 51 (Nieuwkoop and Faber 1967) when the mesenchymal condensation of the femur appears. By stage 52, the femur is procartilaginous, upper leg muscle condensations have appeared, and the mesenchymal condensations of the tibia and fibula have emerged. The femur begins to chondrify at stage 53, when the tibia and fibula are procartilaginous, and the foot is represented only by mesenchymal condensations. Chondrification, ossification, and muscle differentiation sweep proximally to distally until differentiation is complete by stages 59–60. Up until stage 53 amputation through the femur results in the formation of a normal blastema and complete limb regeneration. At stage 53, regenerative capacity begins to decline, starting in the femur region. Amputation through the femur at this stage results in regenerates with malformations or missing structures, but amputation through the tibia–fibula or ankle region still results in normal regeneration. Thereafter, regeneration becomes increasingly hypomorphic in a proximal to distal progression until finally, as the tadpole progresses through metamorphosis, only hypomorphic spikes of cartilage regenerate from any level of amputation (Dent 1962; Overton 1963; Korneluk and Liversage 1984; Muneoka et al. 1986). The quality of regeneration after amputation through regions of ossification is lower that after amputation through the soft tissues of the joints (Wolfe et al. 2000).

The increasing hypomorphism of limb regeneration with advancing developmental stage in *Xenopus* is correlated with changes in the cellular features of the blastema (Korneluk and Liversage 1984; Wolfe et al. 2000). Histolysis is minimal and the blastema cells that appear are fibroblastic in morphology, rather than mesenchymal, so the blastema is often referred to as a "pseudoblastema." There is an increase in basement membrane and dermal tissue invasion under the wound epidermis, AEC thickness declines, and there is decreased sprouting of nerves and blood vessels into the blastema (Wolfe et al. 2000). The fibroblastic cells proliferate and then differentiate into several cartilage nodules that fuse to form a symmetrical cartilage spike of varying length. Muscle is not regenerated.

The loss of regenerative capacity in *Xenopus* limbs appears to be the result of intrinsic changes in the limb cells as they differentiate, not to systemic factors. Transplantation of regeneration-deficient to regeneration-competent limb buds and vice versa does not alter the regenerative capacity of the donor limb bud (Sessions and Bryant 1988; Filoni et al. 1991).

Genes whose expression is enriched at regeneration-competent (stage 53) and regeneration-deficient (stage 59) stages of hindlimb development in *Xenopus* have now been identified by suppression PCR-based subtractive hybridization (King et al. 2003). Array screens of subtracted clones identified several hundred whose expression level was at least twofold higher in unamputated vs amputated stage 53 and 59 limbs. These were sequenced and PCR-based expression analyses were done for 30 randomly selected clones on stage 53 blastema vs stage 59 pseudoblastema. Three distinct categories of expression were identified: genes that were upregulated in both blastema and pseudoblastema, genes that were upregulated in the stage 53 blastema and genes that were upregulated in the stage 59 pseudoblastema. These genes encoded hypothetical proteins, proteins for transcription and translation factors, chromatin proteins, metabolic enzymes, hormones, transport proteins, and structural proteins at percentages from 1.4 (structural proteins) to 4.9 (hypothetical proteins and transcription factors). Most of the clones, however, represented novel genes whose function is unknown (67%).

2.2.3.3
Regeneration-Competent Vs Regeneration-Deficient Species

Comparative analyses can be made between species that differ in their ability to regenerate a tissue or complex structure, for example, between the regeneration-deficient froglets of *Xenopus* or *Rana* and regeneration-competent urodeles such as the newt and axolotl. This kind of comparison has been little exploited and would be advantageous for the study of tissues, such as spinal cord, that are delicate to manipulate surgically in regeneration-competent, but very small *Xenopus* or *Rana* tadpoles.

3 Perspectives

We have learned much about the cellular and molecular biology of regeneration, but are far from knowing enough to translate it into a regenerative medicine. The proliferation kinetics, surface antigen phenotypes, and array of transcription factors for several types of adult stem cells have been partially characterized, enabling their enrichment by FACS after harvest, but the molecular phenotype of other adult stem cells remains uncharacterized. A significant area of research is the identification of the soluble and insoluble signals and their transduction pathways that govern the proliferation and differentiation of regeneration-competent cells. In the case of some cells, such as HSCs, many of these signals are known, but in most cases our information is incomplete.

Two very important findings have been made that have implications for regenerative medicine. The first is that some adult stem cells, such as those from the bone marrow, exhibit pluripotency in chimeric embryo and bone marrow reconstitution assays, though the basis of this pluripotency is not clear. This means that autogeneic stem cells could be harvested from a single, easily accessible source, expanded in culture and transplanted into a lesion site of non-regenerating tissue, with the expectation that they would differentiate into site-specific tissue if we can neutralize inhibitors of regeneration and/or provide those factors that promote regeneration. The second finding is that regeneration-competent cells exist in many tissues of the body that normally respond to injury by scarring. Such cells exist in multiple forms, including stem cells, differentiated cells that can undergo compensatory hyperplasia, and differentiated cells that dedifferentiate to become stem cells. This means that the potential for regeneration exists in most, if not all, tissues of the body and that we can potentially induce their regeneration in situ.

Regeneration by either transplantation or activation in situ of regeneration-competent cells will require solution of the same problem: what chemical and physical signals are required at the lesion site to promote regeneration and what inhibitory factors are present and how do we neutralize them to achieve full functional recovery? To answer this question we need to use animal models that are stronger regenerators than just mice and rats. We can learn much from amphibians, worms such as planaria (Agata et al. 2003), and cnidarians (Holstein et al. 2003), all of which have remarkable powers of regeneration. The most direct experimental strategy to identify the chemical signals that promote regeneration and the factors that inhibit it is to make genomic and proteomic comparisons of tissues that regenerate vs tissues that scar.

The first wave of regenerative medicine will undoubtedly be transplantation of autogeneic stem cells as cell suspensions and aggregates, or seeded into a biomimetic scaffold to make a bioartificial tissue. As a clinical treatment, however, stem cell transplantation will be expensive because of the logistics of harvesting and expanding the cells, as well as the procedures required to transplant them. The wave of the future is the chemical induction of regeneration by regeneration-competent cells by insertion of regeneration-promoting genes into cells or the delivery of regeneration-promoting proteins to a lesion site by injection or other means. These will be not only be clinically simple treatments, but will be relatively inexpensive.

How far we will be able to go in our quest for regeneration is anyone's guess. Twenty years ago paralyzed patients were told that there was no hope for recovery from a devastating spinal cord injury. Like other dogmas, this one too has crumbled under the weight of scientific advances, and spinal cord regeneration that restores some degree of function will be a reality within a few years. Will we be able to induce the regeneration of appendages? Hearts, lungs, kidneys? These will be more complex and difficult tasks, but we should not view them as impossible. They will just take a little longer.

References

Aebischer P, Salessiotis AN, Winn SR (1989) Basic fibroblast growth factor released from synthetic guidance channels facilitates peripheral nerve regeneration across long nerve gaps. J Neurosci Res 23:282–289

Adams C, Watt FM (1990) Changes in keratinocyte adhesion during terminal differentiation reduction in fibronectin binding precedes alpha$_5$beta$_1$ integrin loss from the cell surface. Cell 63:425–435

Aguayo AJ (1985) Axonal regeneration from injured neurons in the adult mammalian central nervous system. In: CW Cotman (ed) Synaptic plasticity. Guilford Press, New York, pp 457–483

Ahmad I, Tang L, Pham H (2000) Identification of neural progenitors in the adult mammalian eye. Biochem Biophys Res Commun 270:517–521

Aiken SW, Badylak SF, Toombs JP, Shelbourne KD, Hiles MC, Lantz GC, Van Sickle D (1994) Small intestine submucosa as an intra-articular ligamentous graft material: a pilot study in dogs. VCOT 7:124–128

Akiyama M, Smith LT, Holbrook KA (1996) Growth factor and growth factor receptor localization in the hair follicle bulge and associated tissue in human fetus. J Invest Dermatol 106:391–396

Alaish SM, Yager D, Diegelmann RF, Cohen IK (1994) Hyaluronate receptor expression in fetal fibroblasts. J Pediatr Surg 29:1040–1043

Alison MR, Golding MH, Sarraf CE (1996) Pluripotential liver stem cells. Facultative stem cells located in the biliary tree. Cell Prolif 29:373–402

Alison MR, Poulsom R, Jeffery R, Dhillon P, Quaglia A, Jacob J, Novelli M, Prentice G, Williamson J, Wright NA (2000) Hepatocytes from non-hepatic adult stem cells. Nature 406:257

Allen RE, Dodson MV, Luiten LS (1984) Regulation of skeletal muscle satellite cell proliferation by bovine pituitary fibroblast growth factor. Exp Cell Res 152:154–160

Allbrook D (1981) Skeletal muscle regeneration. Muscle Nerve 4:234–245

Altman J (1962) Are new neurons formed in the brains of adult mammals? Science 135:1127–1128

Altman J (1963) Autoradiographic investigation of cell proliferation in the brains of rats and cats. Anat Rec 145:573–591

Altmann CR, Brivanlou AH (2001) Neural patterning in the vertebrate embryo. Int Rev Cytol 203:447–482

Anderson JE, Liu L, Kardami E (1991) Distinctive patterns of basic fibroblast growth factor (bFGF) distribution in degenerating and regenerating areas of dystrophic (mdx) striated muscles. Dev Biol 147:96–109

Anton ES, Sandrock AW, Matthew WD (1994) Merosin promotes neurite outgrowth and Schwann cell migration in vitro: evidence using an antibody to merosin, ARM-1. Dev Biol 164:133–146

Anversa P, Nadal-Ginard B (2002) Myocyte renewal and ventricular remodeling. Nature 415:240–243

Arsanto J-P, Komorowski TE, Dupin F, Caubit X, Diano M, Geraudie J, Carlson BM, Thouveny Y (1992) Formation of the peripheral nervous system during tail regeneration in urodele amphibians: ultrastructure and immunohistochemical studies of the origin of the cells. J Exp Zool 264:273–292

Atkins BZ, Hueman MT, Meuchel JM, Cottman MJ, Hutcheson KA, Taylor DA (2000) Myocardial cell transplantation improves in vivo regional performance in infarcted rabbit myocardium. Cardiac Vasc Reg 1:43–53

Attardi DG, Sperry RW (1963) Preferential selection of central pathways by regenerating optic nerve fibers. Exp Neurol 7:46–64

Aubin JE, Liu F (1996) The osteoblast lineage. In: Bilezikian J, Raisz LG, Rodan GA (eds) Principles of bone biology. San Diego Academic Press, San Diego, pp 51–68

Azizi SA, Stokes D, Augelli BJ, DiGirolamo C, Prockop DJ (1998) Engraftment and migration of human bone marrow stromal cells implanted in the brains of albino rats—similarities to astrocyte grafts. Proc Natl Acad Sci USA 95:3908–3913

Bader D, Oberpriller JO (1978) Repair and reorganization of minced cardiac muscle in the adult newt (*Notophthalmus viridescens*). J Morph 155:349–358

Badylak SF (2002) The extracellular matrix as a scaffold for tissue reconstruction. Semin Cell Dev Biol 13:377–383

Badylak SF, Coffey AC, Lantz GC, Tacker WA, Geddes LA (1994) Comparison of the resistance to infection of intestinal submucosa arteria autografts versus polytetrafluoroethylene arterial prostheses in a dog model. J Vasc Surg 19:465–472

Baldwin SP, Saltzman WM (1996) Polymers for tissue engineering. Trends Polymer Sci 4:1–7

Barron KD (1983) Axon reaction and central nervous system regeneration. In: Seil FJ (ed) Nerve, organ and tissue regeneration: research perspectives. Academic Press, New York, pp 3–36

Becker CG, Becker T, Meyer RL, Schachner M (1999) Tenascin-R inhibits the growth of optic fibers in vitro but is rapidly eliminated during nerve regeneration in the salamander *Pleurodeles waltl*. J Neurosci 19:813–827

Bellamkonda R, Aebischer P (1995) Tissue engineering in the nervous system. In: Bronzino JD (ed) The biomedical engineering handbook. CRC Press, Boca Raton, pp 1754–1773

Beltrami AP, Urbanek K, Kajstura J, Yan S-M, Finato N, Bussani R, Nadal-Ginard B, Silvestri F, Leri A, Beltrami A, Anversa P (2001) Evidence that human cardiac myocytes divide after myocardial infarction. New Eng J Med 344:1750–1756

Benraiss A, Caubit X, Coulon J, Nicolas S, Le Parco Y, Thouveny Y (1996) Clonal cell cultures from adult spinal cord of the amphibian urodele *Pleurodeles waltl* to study the identity and potentialities of cells during tail regeneration. Dev Dynam 205:135–149

Benraiss A, Arsanto JP, Coulon J, Thouveny Y (1997) Neural crest-like cells originate from the spinal cord during tail regeneration in adult amphibian urodeles. Dev Dynam 209:15–28

Benraiss A, Arsanto JP, Coulon J, Thouveny Y (1999) Neurogenesis during caudal spinal cord regeneration in adult newts. Dev Genes Evol 209:363–369

Berardi A, Wang A, Levine JD, Lopez P, Scadden D (1995) Functional isolation and characterization of human hematopoietic stem cells. Science 267:104–108

Birling M-C, Price J (1995) Influence of growth factors on neuronal differentiation. Curr Opinion Cell Biol 7:878–884

Bischoff R (1986) A satellite cell mitogen from crushed adult muscle. Dev Biol 115:140–147

Bittira B, Wang J-S, Shum-Tim D, Chiu RC-J (2000) Marrow stromal cells as the autologous, adult stem cell source for cardiac myogenesis. Cardiac Vasc Reg 1:205–210

Bjorklund A, Lindvall O (2000) Cell replacement therapies for central nervous system disorders. Nature Neurosci 3:537–544

Bjornson R, Rietze R, Reynolds BA, Magli MC, Vescovi AL (1999) Turning brain into blood: a hematopoietic fate adopted by adult neural stem cells in vivo. Science 283:534–537

Blau H, Pavlath GK, Hardeman E, Chiu C-P, Silberstein L, Webster SG, Miller SC, Webster C (1985) Plasticity of the differentiated state. Science 230:758–766

Blau HM, Brazelton TR, Weimann JM (2001) The evolving concept of a stem cell: entity or function. Cell 105:829–841

Blaveri K, Heslop L, Yu DS, Rosenblatt D, Gross JG, Partridge A, Morgan JE (1999) Patterns of repair of dystrophic mouse muscle: studies on isolated fibers. Dev Dyn 216:244–256

Block GD, Locker WC, Bowen WC, Petersen BE, Katyal S, Strom SC, Tiley T, Howard TA, Michalopoulos GK (1996) Population expansion, clonal growth, and specific differentiation patterns in primary cultures of hepatocytes induced by HGF/SF, EGF and TGF alpha in a chemically defined (HGM) medium. J Cell Biol 132:1133–1149

Bodega G, Suarez I, Rubio M, Fernandez B (1994) Ependyma: Phylogenetic evolution of glial fibrillary acidic protein (GFAP) and vimentin expression in vertebrate spinal cord. Histochem 102:113–122

Bohn R, Reier PJ, Sourbeer EB (1982) Axonal interactions with connective tissue and glial substrata during optic nerve regeneration in *Xenopus* larvae and adults. Am J Anat 165:397–419

Bolander ME (1992) Regulation of fracture repair by growth factors. Proc Soc Exp Biol Med 200:165–170

Bonner-Weir S, Baxter LA, Schuppin GT, Smith FE (1993) A second pathway for regeneration of adult exocrine and endocrine pancreas: A possible recapitulation of embryonic development. Diabetes 42:1715–1720

Borisov AB (1998) Cellular mechanisms of myocardial regeneration. In: Cellular and molecular basis of regeneration. Ferretti P, Geraudie J (eds) John Wiley and Sons, New York, pp 335–354

Borisov AB (1999) Regeneration of skeletal and cardiac muscle in mammals: do non-primate models resemble human pathology? Wound Rep Reg 7:26–35

Bostrom MPG (1998) Expression of bone morphogenetic proteins in fracture healing. Clin Orthopaed Related Res 355S:S116–S123

Boyce ST (1996) Cultured skin substitutes: a review. Tissue Eng 2:255–266

Bradbury EJ, Moon LDF, Popat RJ, King VR, Bennet GS, Patel PN, Fawcett JW, McMahon SB (2002) Chondroitinase ABC promotes functional recovery after spinal cord injury. Nature 416:636–640

Bracken MB, Shepard MJ, Collins WF, Holford TR, Young W, Baskin DS, Eisenberg HM, Flamm E, Leo-Summers L, Maroon JC, Marshall LF, Perot DL, Piepmeier J, Sonntag E, Wagner FC, Wilberger JL, Winn HR, Young W (1990) A randomized controlled trial of methylprednisolone or naloxone in the treatment of acute spinal cord injury. New Eng J Med 322:1405–1461

Bracken MB, Shepard MJ, Collins WF, Holford TR, Young W, Baskin DS, Eisenberg HM, Flamm E, Leo-Summers L, Maroon JC, Marshall LF, Perot DL, Piepmeier J, Sonntag E, Wagner FC, Wilberger JL, Winn HR, Young W (1992) Methylprednisolone or naloxone treatment after acute spinal cord injury: 1-year follow-up data. J Neurosurg 76:23–31

Brazelton TR, Rossi FMV, Keshet G, Blau HM (2000) From marrow to brain: expression of neuronal phenotypes in adult mice. Science 290:1775–1779

Briscoe J, Ericson J (2001) Specification of neuronal fates in the ventral neural tube. Curr Opin Neurobiol 11:43–49

Brighton CT, Hunt RM (1991) Early histological and ultrastructural changes in medullary fracture callus. J Bone Joint Surg 73A:832–847

Brittberg M, Lindahl A, Nilsson AA, Ohlsson C, Isaksson O, Peterson L (1994) Treatment of deep cartilage defects in the knee with autologous chondrocyte transplantation. New Engl J Med 331:899–895

Brockes JP (1997) Amphibian limb regeneration: rebuilding a complex structure. Science 276:81–87

Brockes JP, Kumar A (2002) Plasticity and reprogramming of differentiated cells in amphibian regeneration. Nature Reviews: Mol Cell Biol 3:566–574

Brook FA, Gardner RL (1997) The origin and efficient derivation of embryonic stem cells in the mouse. Proc Natl Acad Sci USA 94:5709–5712

Broxmeyer HE, Williams DE (1988) The production of myeloid blood cells and their regulation during health and disease. Crit Rev Oncol Hematol 8:173–226

Bruder SP, Kraus H, Goldberg VM, Kadiyala S (1988) The effects of implants loaded with autologous mesenchymal stem cells on the healing of canine segmental bone defects. J Bone Joint Surg 80:985–996

Bruder SP, Kurth AA, Shea M, Hayes WC, Jaiswal N, Kadiyala S (1998) Bone regeneration by implantation of purified, culture-expanded human mesenchymal stem cells. J Orthop Res 16:155–162

Brustle O, Jones N, Learish R, Karram K, Choudhary K, Wiestler OD, Duncan ID, McKay RDG (1999) Embryonic stem cell-derived glial precursors. A source of myelinating transplants. Science 285:754–756

Bucher NLR, Malt RA (1971) Regeneration of the liver and kidney. Little, Brown and Co, Boston

Bunge R (1987) Tissue culture observations relevant to the study of axon-Schwann cell interactions during peripheral nerve development and repair. J Exp Biol 132:21–34

Butler EG, Ward MB (1965) Reconstitution of the spinal cord following ablation in urodele larvae. J Exp Zool 160:47–66

Butler EG, Ward MB (1967) Reconstitution of the spinal cord following ablation in adult *Triturus*. Dev Biol 15:464–486

Byrne JA, Simonsson S, Gurdon JB (2002) From intestine to muscle: nuclear reprogramming through defective cloned embryos. Proc Natl Acad Sci 99:6059–6063

Cai D, Shen Y, DeBellard M, Tang S, Filbin MT (1999) Prior exposure to neurotrophins blocks inhibition of axonal regeneration by MAG and myelin via a cAMP-dependent mechanism. Neuron 22:89–101

Campbell CJ (1969) The healing of cartilage defects. Clin Orthop 64:45–63

Caplan AI (1991) Mesenchymal stem cells. J Orthop Res 9:641–650

Caplan AI, Fink DJ, Goto T, Linton AE, Young RG, Wakitani S, Goldberg VM, Haynesworth SE (1993) Mesenchymal stem cells and tissue repair. In: Jackson DW, Arnoczky S, Woo S, Frank

C (eds) The anterior cruciate ligament: current and future concepts. Raven Press, New York, pp 405–417

Carlson BM (1983) The biology of muscle transplantation. In: Muscle transplantation. Freilinger G, Holle J, Carlson BM (eds) Springer Verlag, New York, pp 3–18

Carlson BM (2003) Muscle regeneration in amphibians and mammals: passing the torch. Dev Dynam 226:167–181

Carlson BM, Gutmann E (1972) Development of contractile properties of minced muscle regenerates in the rat. Exp Neurol 36:239–249

Carlson BM, Gutmann E (1975) Regeneration in free grafts of normal and denervated muscles in the rat: morphology and histochemistry. Anat Rec 183:47–62

Carnac G, Fajas LL, L'honore A, Sardet C, Lamb NJC, Fernandez A (2000) The retinoblastoma-like protein p130 is involved in the determination of reserve cells in differentiating myoblasts. Curr Biol 10:543–546

Caroni P, Schwab ME (1988) Two membrane protein fractions from rat central myelin with inhibitory properties for neurite growth and fibroblast spreading. J Cell Biol 106:1281–1288

Carrino DA, Oron V, Pechak D, Caplan AI (1988) Reinitiation of chondroitin sulphate proteoglycan synthesis in regenerating skeletal muscle. Development 103:641–656

Castro RF, Jackson KA, Goodell MA, Robertson CS, Liu H, Shine HD (2002) Failure of bone marrow cells to transdifferentiate into neural cells in vivo. Science 297:1299

Caubit X, Arsanto J-P, Figarella-Branger D, Thouveny Y (1993) Expression of polysialylated neural cell adhesion molecule (PSA-N-CAM) in developing, adult and regenerating caudal spinal cord of the urodele amphibians. Int J Develop 37:327–336

Caubit X, Riou JF, Coulon J, Arsanto JP, Benraiss A, Boucaut JC, Thouveny Y (1994) Tenascin expression in developing, adult and regenerating caudal spinal chord in the urodele amphibians. Int J Develop Biol 38:661–672

Caubit X, Nicolas S, Shi DL, Le Parco Y (1997a) Reactivation and graded axial expression pattern of *Wnt-10a* gene during early regeneration stages of adult tail in amphibian urodele *Pleurodeles waltl*. Dev Dyn 208:139–148

Caubit X, Nicolas S, Le Parco Y (1997b) Possible roles for *Wnt* genes in growth and axial patterning during regeneration of the tail in urodele amphibians. Dev Dyn 210:1–10

Chambers I, Colby D, Robertson M, Nichols J, Lee S, Tweedie S, Smith A (2003) Functional expression cloning of Nanog, a pluripotency sustaining factor in embryonic stem cells. Cell 113:643–655

Chen G, Birnbaum R, Yablonka-Reuvini Z, Quinn LS (1994) Separation of mouse crushed muscle extract into distinct mitogenic activities by heparin affinity chromatography. J Cell Physiol 160:563–572

Chen D, Ji X, Harris MA, Feng JQ, Karsenty G, Celeste AJ, Rosen V, Mundy GR, Harris SE (1998) Differential roles for bone morphogenetic protein (BMP) receptor type 1B and 1A in differentiation and specification of mesenchymal precursor cells to osteoblast and adipocyte lineages. J Cell Biol 141:295–305

Chen MS, Huber AB, Can der Haar ME, Fran M, Schnell L, Spillmann AA, Christ F, Schwab ME (2000) Nogo-A is a myelin-associated neurite outgrowth inhibitor and an antigen for monoclonal antibody IN-1. Nature 403:434–439

Cheng H, Cao Y, Olson L (1996) Spinal cord repair in adult paraplegic rats: partial restoration of hind limb function. Science 273:510–513

Chernoff EAG (1996) Spinal cord regeneration: a phenomenon unique to urodeles? Int J Dev Biol 40:823–832

Chernoff EAG, Henry LC, Spotts T (1998) An ependymal cell culture system for the study of spinal cord regeneration. Wound Rep Reg 6:435–444

Chernoff EAG, O'Hara CM, Bauerle D, Bowling M (2000) Matrix metalloproteinase production in regenerating axolotl spinal cord. Wound Rep Reg 8:282–291

Chernoff EAG, Stocum DL, Nye HLD, Cameron JA (2003) Urodele spinal cord regeneration and related processes. Dev Dyn 226:280–294

Chiang YH, Silani V, Zhou FC (1996) Morphological differentiation of astroglial progenitor cells from EGF-responsive neurospheres in response to fetal calf serum, basic fibroblast growth factor, and retinol. Cell Transpl 5:179–189

Chiu RCJ, Zibaitis A, Kao RL (1995) Cellular cardiomyoplasty: myocardial regeneration with satellite cell implantation. Ann Thorac Surg 60:12–18

Chopra V, Blewett CJ, Ehrlich HP, Krummel TM (1997) Transition from fetal to adult repair occurring in mouse forelimbs maintained in organ culture. Wound Rep Reg 5:47–51

Chuah MI, Au C (1991) Olfactory Schwann cells are derived frm precursor cells in the olfactory epithelium. J Neuro Sci Res 29:172–180

Cibelli B, Kiessling AA, Cunniff K, Richards C, Lanza RP, West MD (2001) Somatic cell nuclear transfer in humans: pronuclear and early embryonic development. E-biomed J Reg Med 2:25–31

Cibelli JB, Grant KA, Chapman K, Cunniff, Worst T, Green H, Walker SJ, Gutin P, Vilner L, Tabar V, Dominko T, Kane J, Wettstein PJ, Lanza RP, Studer L, Vrana KE, West MD (2002) Parthenogenetic stem cells in nonhuman primates. Science 295:819

Clark RAF (1996) Wound repair: overview and general considerations. In: Clark RAF (ed) The molecular cellular biology of wound repair. Plenum Press, New York, pp 3–50

Clarke D, Johansson C, Wilbertz J, Veress B, Nilsson E, Karlstrom H, Lendahl U, Frisen J (2000) Generalized potential of adult neural stem cells. Science 288:1660–1663

Clarke JDW, Alexander R, Holder N (1988) Regeneration of descending axons in the spinal cord of the axolotl. Neurosci Lett 89:1–6

Clarke JDW, Ferretti P (1998) CNS regeneration in lower vertebrtates. In: Ferretti P, Geraudie J (eds) Cellular and molecular basis of regeneration. John Wiley and Sons, New York, pp 255–272

Claycomb WC (1991) Proliferative potential of the mammalian ventricular cardiac muscle cell. In: Oberpriller JO, Oberpriller JC, Mauro A (eds) The developmental and regenerative potential of cardiac muscle. Harwood, New York, pp 351–363

Claycomb WC (1992) Control of cardiac muscle cell division. Trends Cardiovasc Med 2:231–236

Cooper DKC, Lanza RP (2000) Xeno. Oxford University Press, New York

Condorelli G, Borello U, De Angelis L, Latronico M, Sirabella D, Coletta M, Galli R, Balconi G, Follenzi A, Frati G, Cusella De Angelis MG, Gioglio L, Amuchastegui S, Adorini L, Naldini L, Vescovi A, Dejana E, Cossu G (2001) Cardiomyocytes induce endothelial cells to trans-differentiate into cardiac muscle: implications for myocardium regeneration. Proc Natl Acad Sci USA 98:10733–10738

Constanz BR, Ison IC, Fulmer MT, Poser RD, Smith ST, VanWagoner M, Ross J, Goldstein SA, Jupiter JB, Rosenthal DI (1995) Skeletal repair by in situ formation of the mineral phase of bone. Science 267:1796–1799

Corcoran S, Maden M (2002) Absence of retinoids can induce motoneuron disease in the adult rat and a retinoid defect is present in motoneuron disease patients. J Cell Sci 115:4735–4741

Cornelison DDW, Wold BJ (1997) Single-cell analysis of regulatory gene expression in quiescent and activated mouse skeletal muscle satellite cells. Dev Biol 191:270–283

Cornelison DDW, Filla MS, Stanley H, Rapraeger AC, Olwin BB (2001) Syndecan-3 and syndecan-4 specifically mark skeletal muscle satellite cells and are implicated in satellite cell maintenance and muscle regeneration. Dev Biol 239:79–94

Costarelis G, Sun T-T, Lavker RM (1990) Label-retaining cells reside in the bulge area of pilosebaceous unit: implications for follicular stem cells, hair cycle and skin carcinogenesis. Cell 61:1329–1337

Cowin A, Brtosnan MP, Holmes TM, Ferguson MWJ (1998) Endogenous inflammatory response to dermal wound healing in the fetal and adult mouse. Dev Dyn 212:385–393

Crowe R, Zikherman J, Niswander L (1999) Delta-1 negatively regulates the transition from prehypertrophic to hypertrophic chondrocytes during cartilage formation. Development 126:987–998

Danielpour D, Sporn MB (1990) Differential inhibition of transforming growth factor $\beta1$ and $\beta2$ activity by α_2 macroglobulin. J Biol Chem 265:6973–6977

Davis BM, Duffy MT, Simpson SB Jr (1989) Bulbospinal and intraspinal connection in normal and regenerated salamander spinal cord. Exp Neurol 103:41–51

Davis BM, Ayers JL, Koran L, Carlson J, Anderson MC, Simpson SB (1990) Time course of salamander spinal cord regeneration and recovery of swimming: HRP retrograde pathway tracing and kinematic analysis. Exp Neurol 108:198–213

DePalma RL, Krummel TM, Durham LA, Michna BA, Thomas BL, Nelson JM, Diegelmann RF (1989) Characterization and quantitation of wound matrix in the fetal rabbit. Matrix 9:224–231

Dennis JE, Solchaga LA, Caplan AI (2001) Mesenchymal stem cells for musculoskeletal tissue engineering. Landes Biosci 1:112–115

Dent JN (1962) Limb regeneration in larvae and metamorphosing individuals of the South African clawed toad. J Morph 110:61–77

Desquenne-Clark L, Clark HK, Heber-Katz E (1998) A new murine model for mammalian wound repair and regeneration. Clin Immunol Immunopathol 88:35–45

DeVries HJC, Middelkoop E, Mekkes JR, Dutrieux R, Wildevuur CHR, Westerhof W (1994) Dermal regeneration in native non-cross-linked collagen sponges with different extracellular matrix molecules. Wound Rep Reg 2:37–47

DiMario J, Buffinger N, Yamada S, Strohman RC (1989) Fibroblast growth factor in the extra cellular matrix of dystrophic (mdx) mouse muscle. Science 244:688–690

Dix DJ, Eisenberg BR (1983) Distribution of myosin mRNA during development and regeneration of skeletal muscle fibers. Dev Biol 143:422–426

Doetsch F, Caille I, Lim DA, Garcia-Verdugo JM, Alvarez-Bullya A (1999) Subventricular zone astrocytes are neural stem cells in the adult mammalian brain. Cell 97:703–716

Draper BK, Komurasaki T, Davidson MK, Nanney LB (2003) Topical epiregulin enhances repair of murine excisional wounds. Wound Rep Reg 11:188–197

DuCros DL, LeBaron RG, Couchman JR (1995) Association of versican with dermal matrices and its potential role in hair follicle development and cycling. J Invest Dermatol 105:426–431

Dudley AT, Ros MA, Tabin CJ (2002) A re-examination of proximodistal patterning during vertebrate limb development. Nature 418:539–544

Duffy MT, Liebich DR, Garner LK, Hawrych A, Simpson SB Jr, David BM (1992) Axonal sprouting and frank regeneration in the lizard tail spinal cord: correlation between changes in synaptic circuitry and axonal growth. J Comp Neurol 316:363–374

Dugan L, Turetsky DM, Du C, Lobner D, Wheeler M, Almli CR, Clifton K-FS, Luh T-Y, Choi DW, Lin T-S (1997) Carboxyfullerenes as neuroprotective agents. Proc Natl Acad Sci USA 94:9434–9439

Duttaroy A, Bourbeau D, Wang E (1998) Apoptosis rate can be accelerated or decelerated by over-expression or reduction of the level of elongation factor EF-1α. Exp Cell Res 238:168–176

Echeverri K, Clarke DW, Tanaka EM (2001) In vivo imaging indicates muscle fiber dedifferentiation is a major contributor to the regenerating tail blastema. Dev Biol 236:151–164

Echeverri K, Tanaka EM (2002) Ectoderm to mesoderm lineage switching during axolotl tail regeneration. Science 298:1993–1996

Egar M, Singer M (1972) The role of ependyma in spinal cord regeneration in the urodele, Triturus. Exp Neurol 37:422–430

Egar M, Singer M (1981) The role of ependyma in spinal cord regrowth. In: Becker RO (ed) Mechanisms of growth control. Charles Thomas Publisher, Springfield, pp 93–106

Eguchi G (1998) Transdifferentiation of as a basis of eye lens regeneration. In: Ferretti P, Geraudie J (eds) Cellular and molecular basis of regeneration. John Wiley and Sons, New York, pp 207–229

Ehrlich HP (1988) The modulation of contraction of fibroblast populated collagen lattices by types I, II and III collagen. Tiss Cell 20:47–50

Einhorn TA (1998) The cell and molecular biology of fracture healing. Clin Orthopaed Related Res 355S:7–21

Elias A, Zheng T, Einarsson O, Landry M, Trow T, Rebert N, Panuska J (1994) Epithelial interleukin-11. Regulation by cytokines, respiratory syncytial virus and retinoic acid. J Biol Chem 269:22261–22268

Ellis I, Banyard J, Schor SL (1997) Differential response of fetal and adult fibroblasts to cytokines: cell migration and hyaluronan synthesis. Development 124:1593–1600

Eng LF, Reier PJ, Houle JD (1987) Astrocyte activation and fibrous gliosis: glial fibrillary acidic protein immunostaining of astrocytes following intraspinal cord grafting of fetal CNS tissue. Prog Brain Res 71:439–455

Eppenberger ME, Hauser J, Baechi T, Schaub M, Brunner UT, Dechesne CA, Eppenberger HM (1988) Immunocytochemical analysis of the regeneration of myofibrils in long-term cultures of adult cardiomyocytes of the rat. Dev Biol 130:1–15

Epstein EH (1974) [Alpha 1 (III)] human skin collagen. J Biol Chem 249:3225–3231

Eriksson PS, Perfilieva E, Bjork-Eriksson T, Alborn AM, Nordberg C, Peterson DA, Gage FH (1998) Neurogenesis in the adult hippocampus. Nature Med 4:1313–1317

Espejo EF, Montoro RJ, Armengol JA, Lopez-Barneo J (1998) Cellular and functional recovery of Parkinsonian rats after intrastriatal transplantation of carotid body cell aggregates. Neuron 20:197–206

Fallon J, Reid S, Kinyamy R, Opole I, Opole R, Baratta J, Korc M, Endo TL, Duong A, Nguyen G, Karkehabadhi M, Twardzik D, Loughlin S (2000) In vivo induction of massive proliferation, directed migration, and differentiation of neural cells in the adult mammalian brain. Proc Natl Acad Sci USA 97:14686–14691

Fausto N (1994) Liver stem cells. In: Arias IM, Boyer JL, Fausto N, Jakoby WB, Schacter DA, Shafritz DA (eds) Raven Press, New York, pp 1501–1518

Fausto N (2000) Liver regeneration. J Hepatol 32:19–31

Fausto N, Webber EM (1994) Liver regeneration. In: Arias IM, Boyer JL, Fausto N, Jakoby WB, Schachter DA, Shafritz DA (eds) The liver biology and pathobiology, 3rd edn. Raven Press, New York, pp 1059–1084

Fausto N, Laird AD, Webber EM (1995) Role of growth factors and cytokines in hepatic regeneration. FASEB J 9:1527–1536

Fawcett JW, Asher RA (1999) The glial scar and central nervous system repair. Brain Res Bull 49:377–391

Feldblum S, Arnaud M, Simon M, Rabin O, d'Arbigny P (2000) Efficacy of a new neuroprotective agent, gacyclidine, in a model of rat spinal cord injury. J Neurotrauma 17:1079–1093

Ferguson CM, Miclau T, Hu D, Alpern E, Helms JA (1998) Common molecular pathways in skeletal morphogenesis and repair. Ann NY Acad Sci 857:33–42

Ferretti P, Zhang F, O'eill P (2003) Changes in spinal cord regenerative ability through phylogenesis and development. Dev Dyn 226:245–256

Ferrari G, Cusella-De Angelis G, Coletta M, Paolucci E, Stornaiuolo A, Cossu G, Mavilio F (1998) Muscle regeneration by bone marrow-derived myogenic progenitors. Science 279:1528–1530

Filbin MT (2000) Axon regeneration: vaccinating against spinal cord injury. Curr Biol 10:R100–R103

Filoni S, Bernardini S, Cannata SM (1991) The influence of denervation on grafted hindlimb regeneration of larval *Xenopus laevis*. J Exp Zool 260:210–219

Fischer AJ, Seltner RLP, Poon J, Stell WK (1998) Immunocytochemical characterization of NMDA and QA-induced excitotoxicity in the retina of chicks. J Comp Neurol 393:1–15

Fischer AJ, Reh TA (2001) Muller glia are a potential source of neural regeneration in the post-natal chicken retina. Nat Neurosci 4:247–252

Fischer AJ, Reh TA (2002) Exogenous growth factors stimulate the regeneration of ganglion cells in the chicken retina. Dev Biol 251:367–379

Fischer A, Dierks BD, McGuire C, Reh TA (2002) Insulin and FGF2 activate a neurogenic program in Muller glia of the chicken retina. J Neurosci 22:9387–9398

Fleet JC, Cashman K, Cox K, Rosen V (1997) The effects of aging on the bone inductive activity of recombinant human bone morphogenetic protein-2. Endocrinology 137:4605–4610

Frantz FW, Diegelmann RF, Mast BA, Cohen KI (1992) Biology of fetal wound healing: collagen biosynthesis during dermal repair. J Pediatr Surg 27:945–949

Frantz FW, Bettinger DA, Haynes JH, Johnson DE, Harvey KM, Dalton HP, Yager DR, Diegelmann RF, Cohen IK (1993) Biology of fetal repair: the presence of bacteria in fetal wounds induces an adult-like healing response. J Pediatr Surg 3:428–434

Friedenstein AJ, Chailakhjan RK, Lalykina KS (1970) The development of fibroblast colonies in monolayer cultures of guinea pig bone marrow and spleen cells. Cell Tissue Kinet 3:393–402

Friedenstein AJ, Chailakhyan RK, Gerasimov UV (1987) Bone marrow osteogenic stem cells: in vitro cultivation and transplantation in diffusion chambers. Cell Tissue Kinet 20:263–272

Fuchs E, Segre JA (2000) Stem cells: a new lease on life. Cell 100:143–156

Fujio K, Evarts RP, Hu Z, Marsden R, Thorgeirsson SS (1994) Expression of stem cell factor and its receptor, c-kit, during liver regeneration from putative stem cells in adult rat. J Lab Invest 70:511–516

Gage FH (2000) Mammalian neural stem cells. Science 287:1433–1438

Garrity PA, Zipursky SL (1995) Neuronal target recognition. Cell 83:177–185

Garrett KL, Anderson JE (1995) Colocalization of bFGF and the myogenic regulatory gene myogenin in dystrophic mdx muscle precursors and young myotubes in vivo. Dev Biol 169:596–608

Garry DJ, Yang Q, Bassel-Duby R, Williams RS (1997) Persistent expression of MNF identifies myogenic stem cells in postnatal muscles. Dev Biol 188:280–294

Gaviria M, Privat A, d'Arbigny P, Kamenka J, Haton H, Ohanna F (2000) Neuroprotective effects of a novel NMDA antagonist, gacyclidine, after experimental contusive spinal cord injury in adult rats. Brain Research 874:200–209

Gaze RM (1959) Regeneration of the optic nerve in *Xenopus laevis*. Quart J Exp Physiol 44:209–308

Gaze M, Grant P (1978) The diencephalic course of regenerating retinotectal fibers in *Xenopus* tadpoles. J Embryol Exp Morph 44:201–216

Gehris AL, Stringa E, Spina J, Desmond ME, Tuan RS, Bennett VD (1997) The region encoded by the alternatively spliced exon IIIA in mesenchymal fibronectin appears essential for chondrogenesis at the level of cellular condensation. Dev Biol 190:191–205

Geraudie J, Nordlander R, Singer M, Singer J (1988) Early stages of spinal ganglion formation during tail regeneration in the newt, *Notophthalmus viridescens*. Am J Anat 183:359–370

Geraudie J, Ferretti P (1998) Gene expression during amphibian limb regeneration. Int J Cytol 180:1–50

Glowacki J (1998) Angiogenesis in fracture repair. Clin Orthopaed Related Res 355S:S82–S89

Go MJ, Eastman DS, Artavanis-Tsakonas S (1998) Cell proliferation control by Notch signaling in *Drosophila* development. Development 125:2031–2040

Goldman SA, Nottebohm F (1983) Neuronal production, migration, and differentiation in a vocal control nucleus of the adult female canary brain. Proc Natl Acad Sci USA 80:2390–2394

Goldsmith HS, de la Torre JC (1992) Axonal regeneration after spinal cord transection and reconstruction. Brain Res 589:217–224

Gonzales S, Grillo C, Deniselle MCG, Lima A, McEwen BS, De Nicola AF (1996) Dexamethasone upregulates mRNA for Na^+, K^+-ATPase in some spinal cord neurons after cord transection. NeuroReport 7:1041–1044

Goodell MA, Brose K, Paradis G, Conner AS, Mulligan RC (1996) Isolation and functional properties of murine hematopoietic stem cells that are replicating in vivo. J Exp Med 183:1797–1806

Goodrum JF, Bouldin TW (1996) The cell biology of myelin degeneration and regeneration in the peripheral nervous system. J Neuropath Exp Neurol 55:943–953

Gorza L, Sartore S, Triban C, Schiaffino S (1983) Embryonic-like myosin heavy chains in regenerating chicken muscle. Exp Cell Res 143:395–403

Gosiewska A, Yi C-F, Brown LJ, Cullen B, Silcock D, Geesin JC (2001) Differential expression and regulation of extracellular matrix-associated genes in fetal and neonatal fibroblasts. Wound Rep Reg 9:213–222

Goss RJ (1969) Principles of regeneration. Academic Press, New York

Goss RJ (1991) The natural history (and mystery) of regeneration. In: Dinsmore CE (ed) A history of regeneration research. Cambridge University Press, New York, pp 7–24

Gould E, Tanapat P (1997) Lesion-induced proliferation of neuronal progenitors in the dentate gyrus of the adult rat. Neurosci 80:427–436

Gould E, Tanapat P, McEwen BS, Flugge G, Fuchs E (1998) Proliferation of granule cell precursors in the dentate gyrus of adult monkeys is diminished by stress. Proc Natl Acad Sci USA 95:3168–3171

Gould E, Reeves AJ, Graziano MSA, Gross CG (1999) Neurogenesis in the neocortex of adult primates. Science 286:548–552

Grande DA, Southerland SS, Manji R, Pate DW, Schwartz RE, Lucas PA (1995) Repair of articular cartilage defects using mesenchymal stem cells. Tissue Eng 1:345–353

Grant P, Tseng Y (1986) Embryonic and regenerating *Xenopus* retinal fibers are intrinsically different. Dev Biol 114:475–491

GrandPre T, Nakamura F, Vartanian T, Strittmatter SM (2000) Identification of the Nogo inhibitor
of axon regeneration as a Reticulon protein. Nature 403:439–443

GrandPre T, Li S, Strittmatter SM (2002) Nogo-66 receptor antagonist peptide promotes axonal
regeneration. Nature 417:547–55

Griesler HQ (2001) Biomolecules. In: Holdridge GM (ed) WTEC panel report on tissue engineering research. International Technology Research Institute, Baltimore, pp 31–47

Griffin JW, Hoffman PN (1993) Degeneration and regeneration in the peripheral nervous system.
In: Dyck PJ, Thomas PK, Griffin JW, Low PA, Poduslo J (eds) Peripheral neuropathy, 3rd edn.
WB Saunders, Philadelphia, pp 361–376

Griffith LG (2000) Polymeric biomaterials. Acta Mater 48:263–277

Grompe M, Finegold M (2001) Liver stem cells. In: Marshak DR, Gardner RL, Gottlieb D (eds)
Stem cell biology. Cold Spring Harbor Laboratory Press, Cold Spring Harbor, pp 455–497

Gross J (1996) Getting to mammalian wound repair and amphibian limb regeneration: a mecha-
nistic link in the early events. Wound Rep Reg 4:190–202

Grounds MD (1991) Towards understanding skeletal muscle regeneration. Pathol Res Pract 187:1–
22

Grounds MD, Yablonka-Reuvini Z (1993) Molecular and cell biology of skeletal muscle regeneration. In: Partridge TA (ed) Molecular and cell biology of muscular dystrophy. Chapman and
Hall, London, pp 210–256

Gu D-L, Sarvetnick N (1993) Epithelial cell proliferation and islet neogenesis in IFN-gamma transgenic mice. Development 118:33–46

Gu D-L, Lee MS, Krahl T, Sarvetnick N (1994) Transitional cells in the regenerating pancreas. Development 120:1873–1881

Geuna S, Borrione P, Fornaro M, Giacobini-Robecchi MG (2001) Adult stem cells and neurogenesis: historical roots and state of the art. Anat Rec 265:132–141

Guenard V, Kleitman N, Morrissey T, Bunge RP, Aebischer P (1992) Syngeneic Schwann cells derived from adult nerves seeded in semipermeable guidance channels enhance peripheral
nerve regeneration. J Neurosci 2:3310–3320

Guerret S, Govignon E, Hartmann DJ, Ronfard V (2003) Long-term remodeling of a bilayered liv-
ing human skin equivalent (Apligraf) grafted onto nude mice: immunolocalization of human
cells and characterization of extracellular matrix. Wound Rep Reg 11:35–45

Gulati AK, Reddi AH, Zalewski AA (1983) Changes in the basement membrane zone components
during skeletal muscle fiber degeneration and regeneration. J Cell Biol 97:957–962

Gussoni E, Soneoka Y, Strickland CD, Buzney EA, Khan MK, Flint AF, Kunkel LM, Mulligan RC
(1999) Dystrophin expression in the mdx mouse restored by stem cell transplantation. Nature
401:390–394

Ham AW, Cormack DH (1978) Histology, 8th edn. JB Lippincott, Philadelphia, pp 614–644

Hansen-Smith FM, Carlson BM (1979) Cellular responses to free grafting of the extensor digitorum longus muscle of the rat. J Neurol Sci 41:149–173

Hardikar AA, Karandikar MS, Bhonde RR (1999) Effect of partial pancreatectomy on diabetic status in BALB/c mice. J Endocrinol 162:189–195

Hardy MH (1992) The secret life of the hair follicle. Trends Genet 8:55–61

Harrison E, Chen J, Astle CM (2001) Repopulating patterns of primitive hematopoietic stem cells.
In: Marshak DR, Gardner RL, Gottleib D (eds) Stem cell biology. Cold Spring Harbor Laboratory Press, Cold Spring Harbor, pp 111–128

Haynes JH, Johnson DE, Flood LC, Mast BA, Habacker TA, Diegelmann RF, Cohen IK, Krummel
TM (1990) Platelet-derived growth factor induces fibrosis at a fetal wound site. Surg Forum
41:641–643

Hebert JM, Rosenquist T, Gotz J, Martin GR (1994) FGF5 as a regulator of the hair growth cycle:
evidence from targeted and spontaneous mutations. Cell 78:1017–1025

Hellstrom S, Laurent C (1987) Hyaluronan and healing of typanic membrane perforations. An experimental study. Acta Otolaryngol (Stockholm) 42 (Suppl):54–61

Henry EW, Chiu TH, Nyilas E, Brushart TM, Dikkes P, Sidman RL (1985) Nerve regeneration
through biodegradable polyester tubes. Exp Neurol 90:652–676

Higgins GM, Anderson RM (1931) Experimental pathology of the liver: I. Restoration of the liver of the white rat following partial surgical removal. Arch Pathol 12:186–202

Hiles MC, Badylak SF, Lantz GC, Kokini K, Geddes LA, Morff RJ (1995) Mechanica–190

Holder N, Clarke JDW, Kamalati T, Lane EB (1990) Heterogeneity in spinal radial glia demonstrated by intermediate filament expression and HRP labeling. J Neurocytol 19:915–928

Holder N, Clarke JDW, Stephens N, Wilson SW, Orsi C, Bloomer T, Tonge DA (1991) Continuous growth of the motor system in the axolotl. J Comp Neurol 303:534–550

Holtzer H (1952) Reconstitution of the urodele spinal cord following unilateral ablation. J Exp Zool 119:263–301

Honda H, Tanemura M, Imayama S (1996) Spontaneous architectural organization of mammalian epidermis from random cell packing. J Invest Dermatol 106:312–315

Hong Y, Winkler C, Schartl M (1998) Efficiency of cell culture derivation from blastula embryos and of chimera formation in the medaka (*Oryzias latipes*) depends on donor genotype and passage number. Dev Genes Evol 208:595–602

Horne KA, Jahoda CAB, Oliver RF (1986) Whisker growth induced by implantation of cultured vibrissa dermal papilla cells in the adult rat. J Embryol Exp Morph 97:111–124

Horner PJ, Gage FH (2000) Regenerating the damaged central nervous system. Nature 407:963–969

Horwitz EM, Prockop DJ, Fitzpatrick L, Koo WWK, Gordon P, Neel M, Sussman M, Orchard P, Marx JC, Pyeritz RE, Brenner MK (1999) Transplantability and therapeutic effects of bone marrow-derived mesenchymal cells in children with osteogenesis imperfecta. Nature Med 5:309–313

Hotchin NA, Gandarillas A, Watt FM (1995) Regulation of cell surface beta$_1$ integrin levels during keratinocyte terminal differentiation. J Cell Biol 128:1209–1219

Houghton PE, Keefer KA, Krummel T (1995) The role of transforming growth factor-β in the conversion from "scarless" healing to healing with scar formation. Wound Rep Reg 3:229–236

Huang DW, McKerracher L, Braun PE, David S (1999) A therapeutic vaccine approach to stimulate axon regeneration in the adult mammalian spinal cord. Neuron 24:639–647

Hubner G, Hu Q, Smola H, Werner S (1996) Strong induction of activin expression after injury suggests an important role of activin in wound repair. Dev Biol 173:490–498

Hubner K, Fuhrmann G, Christenson LK, Kehler J, Reinhold R, De la Fuente R, Wood J, Strauss JF 3rd, Boiani M, Scholer HR (2003) Derivation of oocytes from mouse embryonic stem cells. Science 300:1251–1256

Humpherys D, Eggan K, Akutsu H, Hochedlinger K, Rideout WM III, Biniszkiewicz D, Yanagimachi R, Jaenisch R (2001) Epigenetic instability in ES cells and cloned mice. Science 293:95–97

Hunter K, Tonge DA, Holder N (1988) In vitro neurite outgrowth from axolotl neurons. Neurosci Lett Sup. 32:S68 "Explant Culture"

Hunter K, Maden M, Summerbell D, Eriksson U, Holder N (1991) Retinoic acid stimulates neurite outgrowth in the amphibian spinal cord. Proc Natl Acad Sci USA 88:3666–3670

Hunziker EB, Rosenberg LL (1996) Repair of partial-thickness defects in articular cartilage: cell recruitment from the synovial membrane. J Bone Joint Surg 78A:721–733

Hynes M, Rosenthal A (1999) Specification of dopaminergic and serotoergic neurons in the vertebrate CNS. Curr Opinion Neurobiol 9:26–36

Ihara S, Motobuyashi Y, Nagao E (1992) Wound closure in foetal rat skin. Development 114:573–582

Ikuta K, Kina T, MacNeil I, Uchida N, Peault B, Chien Y-H, Weissman I (1990) A developmental switch in thymic lymphocyte maturation potential occurs at the level of hematopoietic stem cells Cell 62:863–874

Inamatsu M, Matsuzaki T, Iwanari H, Yoshizato K (1998) Establishment of rat dermal papilla cell lines that sustain the potency to induce hair follicles from afollicular skin. J Invest Dermatol 111:767–775

Ivanova NB, Dimos JT, Schaniel C, Hackney JA, Moore KA, Lemischka IR (2002) Stem cell molecular signature. Science 298:601–604

Iwashita Y, Kawaguchi S, Murata M (1994) Restoration of function by replacement of spinal cord segments in the rat. Nature 367:167–170

Jackson KA, Mi T, Goodell MA (1999) Hematopoietic potential of stem cells isolated from murine skeletal muscle. Proc Natl Acad Sci USA 96:14482–14486

Jackson KA, Majika SM, Wang H, Pocius J, Hartley CJ, Majesky MW, Entman ML, Michael LH, Hirschi K, Goodell MA (2001) Regeneration of ischemic cardiac muscle and vascular endothelium by adult stem cells. J Clin Invest 107:1395–1402

Janeway CA, Travers P, Hunt S, Walport M (1997) Immunobiology, 3rd edn. Garland Pub Co, New York

Jennische E, Ekberg S, Matejka GL (1993) Expression of hepatocyte growth factor in growing and regenerating rat skeletal muscle. Am J Physiol 265:122–128

Jensen UB, Lowell S, Watt F (1999) The spatial relationship between stem cells and their progeny in the basal layer of human epidermis: a new view based on whole mount labeling and lineage analysis. Development 126:2409–2418

Jiang H, Hidaka K, Morisaki H, Morisaki T (2000) MS-5 bone marrow stromal cell line differentiates into cardiac muscle cells after long-tern culture. Cardiac Vasc Reg 1:274–282

Jiang Y, Jahagirdar BN, Reinhardt R, Schwarts RE, Keene CD, Ortiz-Gonzalez XR, Reyes M, Lenvik T, Lund T, Blackstad M, Du J, Aldrich S, Lisberg A, Low WC, Largaespada DA, Verfaille CM (2002) Pluripotency of mesenchymal stem cells derived from adult marrow. Nature 418:41–49

Jin Y, Yang LJ (1990) The relationship between bone morphogenetic protein and neoplastic bone diseases. Clin Orthopaed 259:233–238

Jin K, Minami M, Lan JQ, Mao XO, Batteur S, Simon RP, Greenberg DA (2001) Neurogenesis in dentate subgranular zone and rostral subventricular zone after focal cerebral ischemia in the rat Proc Natl Acad Sci USA 98:4710–4715

Jensen UB, Lowell S, Watt F (1999) The spatial relationship between stem cells and their progeny in the basal layer of human epidermis: a new view based on whole mount labeling and lineage analysis. Development 126:2409–2418

Jindo T, Tsuboi R, Imai R, Takamori K, Rubin JS, Ogawa H (1994) Hepatocyte growth factor/scatter factor stimulates hair growth of mouse vibrissae in organ culture. J Invest Dermatol 103:306–309

Jindo T, Tsuboi R, Imai R, Takamori K, Rubin JS, Ogawa H (1994) Hepatocyte growth factor/scatter factor stimulates hair growth of mouse vibrissae in organ culture. J Invest Dermatol 103:306–309

Johansson C, Momma S, Clarke D, Risling M, Lendahl U, Frisen J (1999) Identification of a neural stem cell in the adult mammalian central nervous system. Cell 96:25–34

Johnson SE, Allen RE (1995) Activation of skeletal muscle satellite cells and the role of fibroblast growth factor receptors. Exp Cell Res 219:449–453

Jones PH (1982) In vitro comparison of embryonic myoblasts and myogenic cells isolated from regenerating adult rat skeletal muscle. Exp Cell Res 139:401–404

Jones PH, Watt FM (1993) Separation of human epidermal stem cells from transit amplifying cells on the basis of differences in integrin function and expression. Cell 73:713–724

Jones PH, Harper S, Watt FM (1995) Stem cell patterning and fate in human epidermis. Cell 80:83–93

Kadiyala S, Jaiswal N, Bruder SP (1997) Culture-expanded bone marrow-derived mesenchymal stem cells can regenerate a critical-sized segmental bone defect. Tissue Eng 3:173–185

Kaku DA, Giffard RG, Choi DW (1993) Neuroprotective effects of glutamate antagonists and extracellular acidity. Science 260:1516–1518

Kao RL, Chin TK, Ganote CE, Hossler FE, LI C, Browder W (2000) Satellite cell transplantation to repair injured myocardium. Cardiac and Vasc Reg 1:31–42

Kapfhammer JP, Schwab ME (1992) Modulators of neuronal migration and neurite growth. Curr Opinion Cell Biol 4:863–868

Kataoka K, Suzuki Y, Kitada M, Ohnishi K, Suzuki K, Tanihara M, Ide C, Endo K, Nishimura Y (2001) Alginate, a bioresorbable material derived from brown seaweed, enhances elongation of amputated axons of spinal cord in infant rats. J Biomed Mater Res 54:373–384

Kato T, Honmou O, Uede T, Hashi K, Kocsis JD (2000) Transplantation of human olfactory ensheathing cells elicits remyelination of demyelinated rat spinal cord. Glia 30:209–218

Kehl LJ, Fairbanks CA, Laughlin TM, Wilcox GL (1997) Neurogenesis in postnatal rat spinal cord: a study in primary culture. Science 276:586–589

Kempermann G, Kuhn HG, Gage FH (1998) Experience-induced neurogenesis in the senescent dentate gyrus. J Neurosci 18:3206–3212

Khalyfa A, Carlson BM, Carlson JA, Wang E (1999) Toxin injury-dependent switched expression between EF-1a and its sister, S1, in rat skeletal muscle. Dev Dyn 216:267–273

Kherif S, Lafuma C, Dehaupas M, Lachkar S, Fournier J-G, Verdiere-Sahuque M, Fardeau M, Alameddine HS (1999) Expression of matrix metalloproteinases 2 and 9 in regenerating skeletal muscle: a study in experimentally injured and *mdx* muscles. Dev Biol 205:158–170

Kiffmeyer WR, Tomusk EV, Mescher AL (1991) Axonal transport and release of transferrin in nerves of regenerating amphibian limbs. Dev Biol 147:392–402

Kim J-H, Auerbach J, Rodriguez-Gomez JA, Velasco I, Gavin D, Lumelsky N, Lee S-H, Nguyen J, Sanchez-Pernaute R, Bankiewicz K, McKay R (2002) Dopamine neurons derived from embryonic stem cells function in an animal model of Parkinson's disease. Nature 418:50–56

King MW, Nguyen T, Calley J, Harty MW, Muzinich C, Mescher AL, Chalfant C, N'Cho MN, McLeaster K, McEntire J, Stocum DL, Smith RC, Neff AW (2003) Identification of genes expressed during *Xenopus laevis* regeneration by using subtractive hybridization. Dev Dyn 226:398–409

Kingsley DM, Bland AE, Grubber JM, Marker PC, Russel LB, Copeland NG, Jenkins NA (1992) The mouse short ear skeletal morphogenesis locus is associated with defects in a bone morphogenetic member of the TGF beta superfamily. Cell 71:399–410

Kirshenbaum LA, Abdellatif M, Chakraborty S, Schneider MD (1996) Human E2F-1 reactivates cell cycle progression in ventricular myocytes and represses cardiac gene transcription. Dev Biol 179:402–411

Klug M, Soonpaa MH, Field LJ (1995) DNA synthesis and multinucleation in embryonic stem cell-derived cardiomyocytes. Am J Physiol 269:H1913–H1921

Klug M, Soonpa MH, Koh GY, Field (1996) Genetically selected cardiomyocytes from differentiating embryonic stem cells form stable intracardiac grafts. J Clin Invest 98:1–9

Knudson CB, Toole BP (1985) Changes in the pericellular matrix during differentiation of limb bud mesoderm. Dev Biol 112:308–318

Kobayashi K, Nishimura E (1989) Ectopic growth of mouse whiskers from implanted lengths of plucked vibrissa follicles. J Invest Dermatol 92:278–282

Kobayashi K, Rochat A, Barrandon Y (1993) Segregation of keratinocyte colony-forming cells in the bulge of the rat vibrissa. Proc Natl Acad Sci USA 90:7391–7395

Kocher AA, Schuster MD, Szabolcs MJ et al (2001) Neovascularization of ischemic myocardium by human bone-marrow-ferived angioblasts prevents cardiomyocytes apoptosis, reduces remodeling and improves cardiac function. Nature Med 7:430–436

Koh GY, Klug MG, Soonpaa MH, Field LJ (1993) Differentiation and long term survival of C2C12 myoblast grafts in heart. J Clin Invest 91:1548–1554

Koh GY, Soonpaaa MH, Klug M, Pride HP, Cooper BJ, Zipes DP, Field LJ (1995) Stable fetal cardiomyocyte grafts in the hearts of dystrophic mice and dogs. J. Clin Invest 96:2034–2042

Koishi K, Zhang M, McLennan IS, Harris AJ (1995) MyoD protein accumulates in satellite cells and is neurally regulated in regenerating myotubes and skeletal muscle fibers. Dev Dyn 202:244–254

Kopen GC, Prockop DJ, Phinney DG (1999) Marrow stromal cells migrate throughout forebrain and cerebellum where they differentiate into astrocytes after injection into neonatal mouse brains. Proc Natl Acad Sci U SA 96:10711–10716

Kordower JH, Emborg ME, Bloch J, Ma SY, Chu Y, Leventhal L, McBride J, Chen E-Y, Palfi S, Roitberg BZ, Brown WD, Holden JE, Pyzalski R, Taylor MD, Carvey P, Ling ZD, Trono D, Hantraye P, Deglon N, Aebischer P (2000) Neurodegeneration prevented by lentiviral vector delivery of GDNF in primate models of Parkinson's disease. Science 290:767–773

Kornack D, Rakic P (2001) Cell proliferation without neurogenesis in adult primate neocortex. Science 294:2127–2130

Korneluk RG, Liversage RA (1984) Tissue regeneration in the amputated forelimb of *Xenopus laevis* froglets. Canadian J Zool 62:2383–2391

Kratz G, Compton CC (1997) Tissue expression of transforming growth factor-β1 and transforming growth factor-α during wound healing in human skin explants. Wound Rep Reg 5:222–228

Krause DS, Theise ND, Collector MI, Hwang S, Gardner R, Neutzel S, Sharkis J (2001) Multi-organ, multilineage engraftment by a single bone marrow-derived stem cell. Cell 105:369–377

Kritzik MR, Sarvetnick N (2001) Pancreatic stem cells. In: Marshak DR, Gardner RL, Gottlieb D (eds) Stem cell biology. Cold Spring Harbor Laboratory Press, Cold Spring Harbor, pp 499–513

Krummel TM, Nelson JM, Diegelmann RF, Lindblad WJ, Salzberg AM, Greenfield LJ, Cohen IK (1987) Fetal response to injury in the rabbit. J Pediatr Surg 22:640–643

Krummel TM, Michna BA, Thomas BL, Sporn MB, Nelson J, Salzberg AM, Cohen I (1988) Transforming growth factor beta induces fibrosis in a fetal wound model. J Pediatr Surg 23:647–652

Kuffler DP (1987) Long distance regulation of regenerating frog axons. J Exp Biol 132:151–160

Kurek B, Nouri S, Kannourakis G, Murphy M, Austin L (1996) Leukemia inhibitory factor and interleukin-6 are produced by diseased and regenerating skeletal muscle. Muscle Nerve 19:1291–1301

Labarge MA, Blau HM (2002) Biological progression from adult bone marrow to mononucleate stem cell to multinucleate muscle fiber in response to injury. Cell 111:589–601

Lachgar S, Moukadiri H, Jonca F, Charveron M, Bouhaddioui N, Gall Y, Bonafe JL, Plouet J (1996) Vascular endothelial growth factor is an autocrine growth factor for hair dermal papilla cells. J Invest Dermatol 106:17–23

Lagasse E, Connors H, Al-Dhalimy M, Teitsma M, Dohse M, Osborne L, Wang X, Finegold M, Weissman I, Grompe M (2000) Purified hematopoietic stem cells can differentiate to hepatocytes in vivo. Nat Med 6:1229–1234

Lambotte L, Saliez A, Triest S, Maiter D, Baranski A, Barker A, Li B (1997) Effect of sialoadenectomy and epidermal growth factor administration on liver regeneration after partial hepatectomy. Hepatology 25:607–612

Lang DM, Rubin BP, Schwab ME, Stuermer CA (1995) CNS myelin and oligodendrocytes of the *Xenopus* spinal cord—but not optic nerve—are nonpermissive for axon growth. Neurosci 15:99–109

Langer R and Vacanti JP (1992) Tissue engineering. Science 260:920–926

Lantz GL, Badylak SF, Hiles MC, Coffey AC, Geddes LA, Kokini K, Sandusky GE, Morff RJ (1993) Small intestinal submucosa as a vascular graft: a review. J Invest Surg 6:297–310

Laurencin CT, Attawia MA, Elgendy HE, Herbert KM (1996) Tissue engineered bone-regeneration using degradable polymers: the formation of mineralized matrices. Bone 19:93s–99s

Lavker RM, Bertolino AP, Freedberg IM, Sun T-T (1999) Biology of hair follicles. In: Freedberg IM, Eisen Z, Wolff K (eds) Fitzpatrick's dermatology in general medicine, 5th edn. McGraw Hill, New York, pp 230–238

LeCouter J, Moritz DR, Li B, Phillips G, Liang XH, Gerber H-P, Hillan KJ, Ferrara N (2003) Angiogenesis-independent endothelial protection of the liver: role of VEGFR-1 Science 299:890–893

Lee K, Deeds JD, Serge GV (1995) Expression of parathyroid hormone-related peptide and its receptor messenger ribonucleic acids during fetal development in rats. Endocrinology 136:453–463

Lefcourt F, Venstrom K, McDonald JA, Reichardt LF (1992) Regulation of expression of fibronectin and its receptor, $\alpha5\beta1$, during development and regeneration of peripheral nerve. Development 116:767–782

Leferovich J, Bedelbaeva K, Samulewicz S, Zhang X-M, Zwas D, Lankford EB, Heber-Katz E (2001) Heart regeneration in adult MRL mice. Proc Natl Acad Sci USA 98:9830–9835

Leifer D, Lipton S, Barnstable CJ, Masland RH (1984) Monoclonal antibody to Thy-1 enhances regeneration of processes by rat retinal ganglion cells in culture. Science 224:303–306

Lian JB, Stein GS, Canalis E, Robey PG, Boskey A (1999) Bone formation: osteoblast lineage cells, growth factors, matrix proteins, and the mineralization process. In: Favus MJ (ed) Primer on the metabolic bone diseases and disorders of mineral metabolism, 4th edn. Lippincott, Williams, and Wilkens, Philadelphia, pp 14–29

Lin RY, Sullivan KM, Argenta PA, Lorenz HP, Adzick NS (1994) Scarless human fetal skin repair is intrinsic to the fetal fibroblast and occurs in the absence of an inflammatory response. Wound Rep Reg 2:297–305

Linares HA (1996) From wound to scar. Burns 22:339–352

Lindvall O, Hagell P (2001) Cell therapy and transplantation in Parkinson's disease. Clin Chem Lab Med 39:356–361

Liu J, Solway K, Messing RO, Sharp FR (1998) Increased neurogenesis in the dentate gyrus after transient global ischemia in gerbils. J Neurosci 18:7768–7778

Liu S, Qu Y, Stewart J, Howard MJ, Chakraborty S, Holekamp F, McDonald JW (2000) Embryonic stem cells differentiate into oligodendrocytes and myelinate in culture and after spinal cord transplantation. Proc Natl Acad Sci USA 97:6126–6131

Livesey FJ, O'Brien A, Li M, Smith AG, Murphy LJ, Hunt SP (1997) A Schwann cell mitogen accompanying regeneration of motor neurons. Nature 390:614–618

Lois C, Alvarez-Buylla A (1994) Long-distance neuronal migration in the adult mammalian brain. Science 264:1145–1148

Longaker MT, Chiu ES, Adzick NS, Stern RL, Harrison MR (1991) Studies in fetal wound healing. V. A prolonged presence of hyaluronic acid characterizes fetal wound fluid. Ann Surg 213:292–296

Longaker MT, Whitby DJ, Jennings RW, Duncan BW, Ferguson MWJ, Harrison MR, Adzick NS (1991) Fetal diaphragmatic wounds heal with scar formation. J Surg Res 50:375–385

Longaker MT, Adzick NS (1991) The biology of fetal wound healing: a review. Plast Reconstr Surg 87:788–798

Longaker M, Bouhana KS, arrison MR, Danielpour D, Roberts AB, Banda MJ (1992) Wound healing in the fetus: possible role for inflammatory macrophages and transforming growth factor-β isoforms

Lorenz HP, Longaker MT, Perkocha LA, Jenninng RW, Harrison MR, Adzick NS (1992) Scarless wound repair: a human fetal skin model. Development 114:253–259

Lowell S, Jones P, Le Roux I, Dunne J, Watt FM (2000) Stimulation of human epidermal differentiation by Delta-Notch signaling at the boundaries of stem-cell clusters. Curr Biol 10:491–500

Luo J, Borgens R, Shi R (2002) Polyethylene glycol immediately repairs neuronal membranes and inhibits free radical production after acute spinal cord injury. J Neurochem 83:471–480

Mackay AM, Beck SC, Murphy JM, Barry FP, Chichester C, Pittenger MF (1998) Chondrogenic differentiation of cultured human mesenchymal stem cells from marrow. Tissue Eng 4:415–428

Magavi SS, Leavitt BR, Macklis JD (2000) Induction of neurogenesis in the neocortex of adult mice. Nature 405:951–955

Maki T, Monaco AP, Mullon CJ, Solomon BA (1996) Early treatment of diabetes with porcine islets in a bioartificial pancreas. Tissue Eng 2:299–306

Makino S, Fukuda K, Miyoshi S, Konishi F, Kodoma H, Pan J, Sano M, Takahashi T, Hori S, Abe H, Hata J, Umezawa A, Ogawa S (1999) Cardiomyocytes can be generated from marrow stromal cells in vitro. J Clin Invest 103:697–705

Mars M, Liu M-L, Kitson RP, Goldfarb RH, Gabauer MK, Michalopoulos GK (1995) Immediate early detection of urokinase receptor after partial hepatectomy and its implications for initiation of liver regeneration. Hepatology 21:1695–1701

Marshak DR, Gottlieb D, Gardner RL (2001) Introduction to stem cell biology. In: Marshak DR, Gardner RL, Gottlieb D (eds) Stem cell biology. Cold Spring Harbor Laboratory Press, Cold Spring Harbor, pp 1–16

Martin P (1997) Wound healing—aiming for perfect skin regeneration. Science 276:75–81

Martinez-Hernandez A, Amenta P (1995) The extracellular matrix in hepatic regeneration. FASEB J 9:1401–1410

Masinde GL, Li X, Gu W, Davidson H, Mohan S, Baylink D (2001) Identification of wound healing/regeneration quantitative trait loci (QTL) at multiple time points that explain seventy percent of variance in (MRL/MpJ and SJL/J) mice F_2 population. Genome Res 11:2027–2033

Mast BA, Flood LC, Haynes JH, DePalma RL, Cohen I, Diegelmann RF, Krummel TM (1991) Hyaluronic acid is a major component of the matrix of fetal rabbit skin and wounds: implications for healing by regeneration. Matrix 11:63–68

Mast BA, Diegelmann R, Krummel TM, Cohen IK (1992) Scarless wound healing in the mammalian fetus. Surg Gynecol Obstetrics 174:441–451

Mast BA, Nelson JM, Krummel TM (1992) Tissue repair in the mammalian fetus. In: Cohen IK, Diegelmann RF, Lindblad WJ (eds) Wound healing: biochemical and clinical aspects. WB Saunders, Philadelphia, pp 326–341

Mast BA (1992) The skin. In: Cohen IK, Diegelmann RF, Lindblad WJ (eds) Wound healing: biochemical and clinical aspects. WB Saunders, Philadelphia, pp 344–355

Mast BA, Frantz FW, Diegelmann RF, Krummel T, Cohen IK (1995) Hyaluronic acid degradation products induce neovascularization and fibroplasia in fetal rabbit wounds. Wound Rep Reg 3:66–72

Mast BA, Haynes JH, Krummel T, Cohen IK, Diegelmann RF (1997) Ultrastructural analysis of fetal rabbit wounds. Wound Rep Reg 3:243–248

Matsuda RM, Spector DH, Strohman RC (1983) Regenerating adult chicken skeletal muscle and satellite cell cultures express embryonic patterns of myosin and tropomyosin isoforms. Dev Biol 100:478–488

Matsuzaki T, Inamatsu M, Yoshizato K (1996) The upper dermal sheath has a potential to regenerate the hair in the rat follicular epidermis. Differentiation 60:287–297

Matsuzaki T, Yoshizato K (1998) Role of hair papilla cells on induction and regeneration processes of hair follicles. Wound Rep Reg 6:524–530

Matus A (2000) Actin-based plasticity in dendritic spines. Science 290:754–758

Mauro A (1961) Satellite cells of skeletal muscle fibers. J Biophys Biochem Cytol 9:493–495

McBrearty BA, Clark LD, Zhang XM, Blankenhorn EP, Heber-Katz E (1998) Genetic analysis of a mammalian wound-healing trait. Proc Natl Acad Sci USA 95:11792–11797

McCallion RL, Ferguson MWJ (1996) Fetal wound healing and the development of antiscarring therapies for adult wound healing. In: RAF Clark (ed) The molecular and cellular biology of wound repair, 2nd edn. Plenum Press, New York, pp 561–600

McClennan IS, Koishi K (1997) Cellular localization of transforming growth factor-beta 2 and -beta 3 (TGF-β2, TGF-β3) in damaged and regenerating skeletal muscles. Dev Dyn 208:278–289

McCluskey J, Martin P (1995) Analysis of the tissue movements of embryonic wound healing—DiI studies in the limb bud stage mouse embryo. Dev Biol 170:102–114

McDonald JW, Sadowsky C (2002) Spinal cord injury. The Lancet 359:417–425

McDonald JW, Liu XZ, Qu Y, Liu S, Mickey SK, Turetsky D, Gottleib DL, Choi DW (1999) Transplanted embryonic stem cells survive, differentiate, and promote recovery in injured rat spinal cord. Nature Med 12:1410–1412

McGann C, Odelberg SJ, Keating MT (2001) Mammalian myotube dedifferentiation induced by newt regeneration extract. Proc Natl Acad Sci USA 98:13699–13703

McKibbin B (1978) The biology of fracture healing in long bones. J Bone Joint Surg 60B:150–162

McKinney-Freeman SL, Jackson KA, Camargo FD, Ferrari G, Mavilio F, Goodell MA (2002) Muscle-derived hematopoietic stem cells are hematopoietic in origin. Proc Natl Acad Sci USA 99:1341–1346

McQuarrie IG (1983) Role of the axonal cytoskeleton in the regenerating nervous system. In: Seil FJ (ed) Nerve, organ and tissue regeneration: research perspectives. Academic Press, New York, pp 51–88

Megeny LA, Kablar B, Garrett K, Anderson JE, Rudnicki MA (1996) MyoD is required for myogenic stem cell function in adult skeletal muscle. Genes Dev 10:1173–1183

Melchers F, Rolink A (2001) Hematopoietic stem cells: lymphopoiesis and the problem of commitment vs plasticity. In: Marshak D, Gardner RL, Gottlieb D (eds) Stem cell biology. Cold Spring Harbor Laboratory Press, Cold Spring Harbor, pp 307–328

Menasche P (2002) Autologous skeletal myoblast transplantation for ischemic cardiomyopathy: first clinical case. Cardiac Vasc Reg 1:155–156

Mendoza AS, Breipohl W, Miragall F (1982) Cell migration from the chick olfactory placode: a light and electron microscopic study. J Embryol Exp Morph 69:47–59

Merkel JR, DiPaolo BR, Hallock GG, Rice DC (1988) Type I and type II collagen content of healing wounds in fetal and adult rats. Proc Soc Exp Biol Med 187:493–497

Messenger AG, Ellit K, Westgate GE, Gibson WT (1991) Distribution of extracellular matrix molecules in human hair follicles. Ann NY Acad Sci 642:253–262

Messenger AG (1993) The control of hair growth: an overview. J Invest Dermatol (Suppl) 101:4S–9S

Messier B, LeBlond CP, Smart I (1958) Presence of DNA synthesis and mitosis in the brain of young adult mice. Exp Cell Res 14:224–226

Meuli M, Lorenz HP, Hedrick MH, Sullivan KM, Harrison MR, Adzick NS (1995) Scar formation in the fetal alimentary tract. J Pediatr Surg 30:392–395

Mezey E, Chandross KJ, Harta G, Maki RA, McKercher SR (2000) Turning blood into brain: cells bearing neuronal antigens generated in vivo from bone marrow. Science 290:1779–1782

Michalopoulos GK, DeFrances MC (1997) Liver regeneration. Science 276:60–66

Migdalska A, Molineaux G, Demuynck H, Evans S, Ruscetti F, Dexter TM (1991) Growth inhibitory effects of transforming growth factor β in vivo. Growth Factors 4:239–245

Millar SE, Willert K, Salinas PC, Roelink H, Nusse R, Sussman DJ, Barsh GS (1999) WNT signaling in the control of hair growth and structure. Dev Biol 207:133–149

Miller RH, Liuzzi FJ (1986) Regional specialization of the radial glial cells of the adult frog spinal cord. J Neurocytol 15:187–196

Miller EJ, Gay S (1992) Collagen structure and function. In: Cohen IK, Diegelmann RF, Lindblad WJ (eds) Wound healing: biochemical and clinical aspects. WB Saunders, Philadelphia, pp 195–208

Miller DH, Khan OA, Sheremat WA, Blumhardt LD, Rice GPA, Libonati MA, Willmer-Hulme A, Dalton CM, Miszkiel KA, O'Connor PW (2003) A controlled trial of natalizumab for relapsing multiple sclerosis. New Eng J Med 348:15–23

Mitsui K, Tokuzawa Y, Itoh H, Segawa K, Murakami M, Takahashi K, Maruyama M, Maeda M, Yamanaka (2003) The homeoprotein Nanog is required for maintenance of pluripotency in mouse epiblast and ES cells. Cell 113:631–642

Miyazawa K, Shimomura T, Kitamura N (1996) Activation of hepatocyte growth factor in injured tissues is mediated by hepatocyte growth factor activator. J Biol Chem 271:3615–3618

Mocchetti I, Rabin S, Colangelo AM, Whittemore SR, Wrathall JR (1996) Increased basic fibroblast growth factor expression following contusive spinal cord injury. Exp Neurol 141:154–164

Molander H, Olsson Y, Engkuist O, Bowald S, Eriksson I (1989) Regeneration of peripheral nerve through a polyglactin tube. Muscle Nerve 5:54–57

Momma S, Johansson SB, Frisen J (2000) Get to know your stem cells. Curr Opinion Neurobiol 10:45–49

Moore R, Walsh FS (1993) The cell adhesion molecule M-cadherin is specifically expressed in developing and regenerating, but not denervated muscle. Development 117:1409–1420

Moriarty KP, Crombleholme T, Gallivan EK, O'Donnell C (1996) Hyaluronic acid-dependent pericellular matrices in fetal fibroblasts: implication for scar-free wound repair. Wound Rep Reg 4:346–352

Moriya T, Hassan AZ, Young W, Chesler M (1994) Dynamics of extracellular calcium activity following contusion of the rat spinal cord. J Neurotrauma 11:255–263

Morris RJ, Potten CS (1999) Highly persistent label-retaining cells in the hair follicles of mice and their fate following induction of anagen. J Invest Dermatol 112:470–475

Morshead CM, Craig CG, van der Kooy D (1998) In vivo clonal analyses reveal the properties of endogenous neural stem cell proliferation in the adult mammalian forebrain. Development 125:2251–2261

Morshead CM, Benveniste P, Iscove NN, van der Kooy D (2002) Hematopoietic competence is a rare property of neural stem cells that may depend on genetic and epigenetic alterations. Nature Med 8:268–273

Muir GD (1999) Locomotor plasticity after spinal injury in the chick. J Neurotrauma 16:705–711

Nag AC (1991) Reactivity of cardiac muscle cells under traumatic conditions. In: Oberpriller JO, Oberpriller JC, Mauro A (eds) The development and regenerative potential of cardiac muscle. Harwood Academic Publishers, New York, pp 313–331

Nag AC, Healy CJ, Cheng M (1979) DNA synthesis and mitosis in adult amphibian cardiac muscle cells in vitro. Science 205:1281–1282

Nakase T, Nomura S, Yoshikawa H, Hashimoto J, Hirota S, Kitamura Y, Oikawa S, Keiro O, Takaoka K (1994) Transient and localized expression of bone morphogenetic protein 4 messenger RNA during fracture healing. J Bone Mineral Res 9:651–659

Nakatomi H, Kuriu T, Okabe S, Yamamoto S, Hatano O, Kawahara N, Tamura A, Kirino T, Nakafuku M (2002) Regeneration of hippocampal pyramidal neurons after ischemic brain injury by recruitment of endogenous neural progenitors. Cell 110:429–441

Nakayama K, Nakayama K (1998) Cip/Kip cyclin-dependent kinase inhibitors: brakes of the cell cycle engine during development. Bioessays 20:1020–1029

Namba RA, Meuli M, Sullivan KM, Le AX, Adzick NS (1998) Spontaneous repair of superficial defects in articular cartilage in a fetal lamb model. J Bone Joint Surg 80:4–10

Newman SA and Tomasek JJ (1996) Morphogenesis of connective tissues. In: Comper WD (ed) Extracellular matrix, vol 2. Molecular components and interactions. Harwood Academic, Amsterdam, pp 335–369

Nicolas S, Caubit X, Massacrier A, Cau P, Le Parco Y (1999) Two Nkx-3-related genes are expressed in the adult and regenerating central nervous system of the urodele Pleurodeles waltl. Dev Genet 24:319–328

Nicolas S, Massacrier A, Caubit X, Cau P, Le Parco Y (1996) A distal-less-like gene is induced in the regenerating central nervous system of the urodele Pleurodeles waltl. Mech Dev 56:209–220

Nichols J, Zevnik B, Anastassiadis K, Niwa H, Klewe-Nebenius D, Chambers I, Scholer H, Smith A (1998) Formation of pluripotent stem cells in the mammalian embryo depends on the POU transcription factor, Oct-4. Cell 95:379–391

Niederost B, Zimmerman DR, Schwab ME, Bandtlow CE (1999) Bovine CNS myelin contains neurite growth-inhibitory activity associated with chondroitin sulfate proteoglycans. J Neurosci 19:8979–9889

Nieuwkoop P, Faber J (1967) Normal table of *Xenopus laevis* (Daudin): a systematical and chronological survey of the development from the fertilized egg till the end of metamorphosis, 2nd edn. Amsterdam: North Holland Pub Co

Nordlander R, Singer M (1978) The role of ependyma in regeneration of the spinal cord in the urodele amphibian tail. J Comp Neurol 180:349–374

Nwomeh C, Liang H-X, Diegelmann R, Cohen K, Yager D (1998) Dynamics of the matrix metalloproteinases MMP-1 and MMP-8 in acute open human dermal wounds. Wound Rep Reg 6:127–134

Nye HLD, Cameron A, Chernoff EAG, Stocum DL (2003) Regeneration of the urodele limb: a review. Dev Dyn 226:280–294

Oberlander A, Tuan S (1994) Expression and functional involvement of N-cadherin in embryonic limb chondrogenesis. Development 120:177–187

Oberpriller JO, Ferrans VJ, Carroll RJ (1983) DNA synthesis in rat atrial myocytes as a response to left ventricular infarction: an autoradiographic study of enzymatically dissociated myocytes. J Mol Cell Cardiol 15:31–42

Odelberg J, Kollhoff A, Keating MT (2000) Dedifferentiation of mammalian myotubes induced by msx1. Cell 103:1099–1109

Olutoye OO, Cohen IK (1996) Fetal wound healing: an overview. Wound Rep Reg 4:66–74

O'Hara CM, Egar MW, Chernoff EAG (1992) Reorganization of the ependyma during axolotl spinal cord regeneration: changes in intermediate filament and fibronectin expression. Dev Dyn 193:103–115

O'Hara CM, Chernoff EAG (1994) Growth factor modulation of injury-reactive ependymal cell proliferation and migration. Tissue Cell 26:599–611

Orkin SH (2001) Hematopoietic stem cells: Molecular diversification and developmental interrelationships. In: Marshak DR, Gardner RL, Gottleib D (eds) Stem cell biology. Cold Spring Harbor Laboratory Press, Cold Spring Harbor, pp 289–306

Orlic D, Kajstura J, Chimenti S, Jakoniuk I, Anderson SM, Li, Pickel J, McKay R, Nadal-Ginard B, Bodine M, Leri A, Anversa P (2001) Bone marrow cells regenerate infarcted myocardium. Nature 410:701–705

Overton J (1963) Patterns of limb regeneration in *Xenopus laevis*. J Exp Zool 154:153–161

Overturf K, al-Dhalimy M, Ou CN, Finegold M, Grompe M (1997) Serial transplantation reveals the stem-cell-like regenerative potential of adult mouse hepatocytes. Am J Pathol 151:1273–1280

Owen M (1987) Marrow stromal cells. J Cell Sci (Suppl) 10:63–76

Owen M, Friedenstein AJ (1988) Stromal stem cells: marrow derived osteogenic precursors. Ciba Foundation Symposium 136:420–460

Pain B, Clark ME, Shen M, Nakazawa H, Sakuri M, Samarut J, Etches R (1996) Long term in vitro culture and characterization of avian embryonic stem cells with multiple morphogenetic potentialities. Development 122:2339-2384

Parentau NL (2001) Cells. In: Holdridge GM (ed) WTEC panel report on tissue engineering research. International Technology Research Institute, Baltimore, pp 17-30

Parker SB, Eichele G, Zhang P, Rawls A, Sands AT, Bradley A, Olsen EN, Harper JW, Elledge SJ (1995) p53-Independent expression of p21Cip1 in muscle and other terminally differentiating cells. Science 267:1024-1027

Pastoret C, Partridge TA (1998) Muscle regeneration. In: Ferretti P, Geraudie J (eds) Cellular and molecular basis of regeneration. John Wiley and Sons, New York, pp 309-334

Paton A, Nottebohm FN (1984) Neurons generated in the adult brain are recruited into functional circuits. Science 225:1046-1048

Peckham PH, Creasy GH (1992) Neural prostheses: clinical applications of functional electrical stimulation in spinal cord injury. Paraplegia 30:96-101

Peppas NA, Langer R (1994) New challenges in biomaterials. Science 263:1715-1720

Pereiera RF, Halford KW, O'Hara MD, Leeper DB, Sokolov B, Pollard MD, Bagasra O, Prockop DJ (1995) Cultured adherent cells from marrow can serve as long-lasting precursor cells for bone, cartilage, and lung in irradiated mice. Proc Natl Acad Sci USA 92:4857-4861

Pesce M, Gross MK, Scholer H (1998) In line with our ancestors: Oct-4 and the mammalian germ. Bioessays 20:722-732

Petersen BE, Goff J, Greenberger JS, Michalopoulos GK (1998) Hepatic oval cells express the hematopoietic stem cell marker Thy-1 in the rat. Hepatology 27:433-445

Petersen BE, Zajac V, Michalopoulos GK (1998) Hepatic oval cell activation in response to injury following chemically induced periportal or pericentral damage in rats. Hepatology 27:1030-1038

Petersen BE, Bowen WC, Patrene KD, Mars M, Sullvan AK, Murase N, Boggs SS, Greenberger JS, Goff JP (1999) Bone marrow as a potential source of hepatic oval cells. Science 284:1168-1170

Peterson L, Bozza MM, Dorner AJ (1996) Interleukin-11 induces intestinal epithelial cell growth arrest through effects of retinoblastoma protein phosphorylation. Am J Pathol 149:895-902

Pfeilschifter J, D'Souza S, Mundy GR (1986) Transforming growth factor beta is released from resorbing bone and stimulates osteoblast activity. J Bone Miner Res 1 (Suppl 1):294

Pfeilschifter J, Bonewald L, Mundy GR (1987) TGF-beta is released from bone with one or more binding proteins which regulate its activity. J Bone Miner Res 2 (Suppl 1):249

Piatt J (1955) Regeneration of the spinal cord in the salamander. J Exp Zool 129:177-208

Pierce GF (1991) Tissue repair and growth factors. In: Dulbecco R (ed) Encyclopedia of human biology. Academic Press, New York, pp 499-509

Pittinger MF, Mackay AM, Beck SC, Jaiswal RK, Douglas R, Mosca JD, Moorman MA, Simonetti W, Craig S, Marshak DR (1999) Multilineage potential of adult human mesenchymal stem cells. Science 284:143-147

Pittinger MF, Marshak DR (2001) Adult mesenchymal stem cells. In: Marshak DR, Gardner RL, Gottlieb D (eds) Stem cell biology. Cold Spring Harbor Laboratory Press, Cold Spring Harbor, pp 349-373

Pluchino S, Quattrini A, Brambilla E, Gritti A, Salani G, Dina G, Galli R, Del Carro U, Amadio S, Bergami A, Furlan R, Comi G, Vescovi A, Martino G (2003) Injection of adult neurospheres induces recovery in a chronic model of multiple sclerosis. Nature 422:688-695

Polesskaya A, Seale P, Rudnicki MA (2003) Wnt signaling induces the myogenic specification of resident CD45+ adult stem cells during muscle regeneration. Cell 113:841-852

Poole TJ, Finkelstein B, Cox C (2001) The role of FGF and VEGF in angioblast induction and migration during vascular development. Dev Dyn 220 1-17

Ponder KP (1996) Analysis of liver development, regeneration and carcinogenesis by genetic marking studies. FASEB J 10:673-682

Potten CS (1974) The epidermal proliferative unit: the possible role of the central basal cell. Cell Tissue Kinet 7:77-88

Potten CS, Morris RJ (1988) Epithelial stem cells in vivo. J Cell Sci (Suppl) 10:45-62

Prinjha R, Moore SE, Vinson M, Blake S, Morrow R, Christie G, Michalovich D, Simmons DL, Walsh FS (2000) Inhibitor of neurite outgrowth in humans. Nature 403:383-384

Prockop DJ (1997) Marrow stromal cells as stem cells for non-hematopoietic tissues. Science 276:71–74

Quarto R, Thomas D, Liang CT (1995) Bone progenitor cell deficits and age-related decline in bone repair capacity. Calcif Tissue Int 56:123–129

Quiani F, Urbanek K, Beltrami AP, Finato N, Beltrami CA, Nadal-Ginard B, Kajstura J, Leri A, Anversa P (2002) Chimerism of the transplanted heart. New Eng J Med 346:5–15

Ramalho-Santos M, Yoon S, Matsuzaki Y, Mulligan RC, Melton DA (2002) "Stemness": transcriptional profiling of embryonic and adult stem cells. Science 298:597–600

Ramiya V, Maraist M, Arfors KE, Schatz DA, Peck B, Cornelius JG (2000) Reversal of insulin-dependent diabetes using islets generated in vitro from pancreatic stem cells. Nature Med 6:278–282

Ramon-Cueto A, Valverde F (1995) Olfactory bulb ensheathing glia: a unique cell type with axonal growth-promoting properties. Glia 14:163–173

Ramon-Cueto A, Plant GW, Avila J, Bunge MB (2000) Long-distance axonal regeneration in the transected adult rat spinal cord is promoted by olfactory ensheathing glia transplants. J Neurosci 18:3803–3815

Rao MS (1999) Multipotent and restricted precursors in the central nervous system. Anat Rec 257:137–148

Rapraeger AC (2000) Syndecan-regulated receptor signaling. J Cell Biol 149:995–998

Reier PJ (1979) Penetration of grafted astrocytic scars by regenerating optic nerve axons in *Xenopus* tadpoles. Brain Res 164:61–68

Reier PJ, Webster H deF (1974) Regeneration and remyelination of *Xenopus* tadpole optic nerve fibers following transection or crush. J Neurocytol 3:591–618

Repesh LA, Fitzgerald TJ, Furcht LT (1982) Fibronectin involvement in granulation tissue and wound healing in rabbits. J Histochem Cytochem 30:351–358

Reyes M, Lund T, Lenvik T, Aguiar D, Koodie L, Verfaille CM (2001) Purification and ex vivo expansion of postnatal human marrow mesodermal progenitor cells. Blood 98:2615–2625

Reyes M, Dudek A, Jahagirdar B, Koodie L, Marker PH, Verfaille CM (2002) Origin of endothelial progenitors in human postnatal bone marrow. J Clin Invest 109:337–346

Reynolds B, Weiss S (1992) Generation of neurons and astrocytes from isolated cells of the adult mammalian central nervous system. Science 255:1707–1710

Reynolds BA, Weiss S (1996) Clonal and population analyses demonstrate that an EGF-responsive mammalian embryonic CNS precursor is a stem cell. Dev Biol 175:1–13

Rietze RL, Valcanis H, Brooker G, Thomas T, Voss AK, Bartlett PF (2001) Purification of a pluripotent neural stem cell from the adult mouse brain. Nature 412:736–739

Robert B, Sassoon D, Jacq C, Gehring W, Buckingham M (1989) Hox-7, a mouse homeobox gene with a novel pattern of expression during embryogenesis. EMBO J 8:91–100

Robertson TA, Grounds MD, Papadimitriou JM (1992) Elucidation of aspects of murine skeletal muscle regeneration using local and whole body irradiation. J Anat 181:265–276

Rochat A, Kobayashi K, Barrandon Y (1994) Location of stem cells of human hair follicles by clonal analysis. Cell 76:1063–1073

Rogers SL (1982) Muscle spindle formation and differentiation in regenerating rat muscle grafts. Dev Biol 94:265–283

Rogers SL, Letourneau PC, Palm SL, McCarthy J, Furcht LT (1983) Neurite extension by peripheral and central nervous system neurons in response to substratum-bound fibronectin and laminin. Dev Biol 98:212–220

Rosen V, Thies RS (1992) The BMP proteins in bone formation and repair. Trends Genet 8:97–102

Roy NS, Wang S, Jiang L, Kang J, Benraiss A, HarrisonpRestelli C, Fraser R, Couldwell WT, Kawaguchi A, Okano H, Nedergaard M, Goldman SA (2000) In vitro neurogenesis by progenitor cells isolated from the adult human hippocampus. Nat Med 6:271–277

Rudman SM, Philpott MP, Thomas GA, Kealey T (1997) The role of IGF-I in human skin and its appendages: morphogen as well as mitogen? J Invest Dermatol 109:770–777

Rudolph R, VandeBerg J, Ehrlich HP (1992) Wound contraction and scar contracture. In: Cohen IK, Diegelmann RF, Lindblad WJ (eds) Wound healing: biochemical and clinical aspects. WB Saunders, Philadelphia, pp 177–194

Rumyantsev PP (1992) Reproduction of cardiac myocytes developing in vivo and its relationship to processes of differentiation. In: Rumyantsev PP (ed) Growth and hyperplasia of cardiac muscle cells. Harwood Press, New York, pp 70–159

Ryu S, Kodama S, Ryu K, Schoenfeld DA, Faustman DL (2001) Reversal of established autoimmune diabetes by restoration of endogenous β cell function. J Clin Invest 108:63–72

Saito T, Dennis JE, Lennon DP, Young RG, Caplan AI (1995) Myogenic expression of mesenchymal stem cells within myotubes of *mdx* mice in vitro and in vivo. Tissue Eng 1:327–343

Salter RB, Harris DJ, Clements ND (1978) The healing of bone and cartilage in transarticular fractures with continuous passive motion. Orthop Trans 2:77

Saltzman WM (1996) Growth factor delivery in tissue engineering. Mater Sci Res Bull 21:62–65

Sanchez-Ramos J, Song S, Cardozo-Pelaez F, Hazzi C, Stedeford T, Willing A, Freeman TB, Saporta S, Janssen W, Patel N, Cooper DR, Sanberg PR (2000) Adult bone marrow stromal cells differentiate into neural cells in vitro. Exp Neurol 164:247–256

Sandusky GE, Lantz GC, Badylak SF (1995) Healing comparison of small intestine submucosa and ePTFE grafts in the canine carotid artery. J Surg Res 58:415–420

Schnell L, Schwab ME (1990) Axonal regeneration in the rat spinal cord produced by an antibody against myelin-associated neurite growth inhibitors. Nature 343:269–272

Schnell L, Schneider R, Kolbeck R, Barde Y-A, Schwab ME (1994) Neurotrophin-3 enhances sprouting of corticospinal tract during development and after adult spinal cord lesion. Nature 367:170–173

Schwab ME, Kapfhammer JP, Bandtlow CE (1993) Inhibitors of neurite outgrowth. Ann Rev Neurosci 16:565–595

Schwartz M. Belkin M, Harel A, Solomon A, Lavie V, Hadani M, Rachailovich I, Stein-Izsak C (1985) Regenerating fish optic nerves and a regeneration-like response to injured optic nerves of adult rabbits. Science 228:600–603

Schwartz RE, Reyes M, Koodie L, Jiang Y, Blackstad M, Lund T, Lenvik T, Johnson S, Hu W-S, Verfaille M (2002) Multipotent adult progenitor cells from bone marrow differentiate into functional hepatocyte-like cells. J Clin Invest 109:1291–1302

Schwob JE (2002) Neural regeneration and the peripheral olfactory system. Anat Rec 269:33–49

Scott TM, Foote J (1981) A study of degeneration, scar formation and regeneration after section of the optic nerve in the frog, *Rana pipiens*. J Anat 133:213–225

Seale P, Rudnicki MA (2000) A new look at the origin, function, and "stem-cell" status of muscle satellite cells. Dev Biol 218115–124

Seale P, Sabourin LA, Girgis-Gabardo A, Mansouri A, Gruss P, Rudnicki MA (2000) Pax7 is required for the specification of myogenic satellite cells. Cell 102:777–786

Seiberg M, Marthinuss (1995) Clusterin expression within skin correlates with hair growth. Dev Dyn 202:294–301

Seitz A, Aglow E, Heber-Katz E (2002) Recovery from spinal cord injury: a new transection model in the C57Bl/6 mouse. J Neurosci Res 67:337–345

Sell S (2001) Heterogeneity and plasticity of hepatocyte lineage cells. Hepatology 2001; 33:738–750

Sempowski GD, Borrello MA, Blieden T, Barth RK, Phipps R (1995) Fibroblast heterogeneity in the healing wound. Wound Rep Reg 3:120–131

Sessions SK, Bryant V (1988) Evidence that regenerative ability is an intrinsic property of limb cells in *Xenopus*. J Exp Zool 247:39–44

Shah M, Foreman DM Ferguson MW (1992) Control of scarring in adult wounds by neutralising antibody to transforming growth factor-beta. Lancet 339:213–214

Shah M, Foreman DM, Ferguson MWJ (1994) Neutralizing antibody to TGF-β1,2 reduces cutaneous scarring in adult rodents. J Cell Sci 107:1137–1157

Shamblott MJ, Axelman J, Wang S, Bugg EM, Littlefield JW, Donovan PJ, Blumenthal D, Huggins GR, Gearhart JD (1998) Derivation of pluripotent stem cells from cultured human primordial germ cells. Proc Natl Acad Sci USA 95:13726–13731

Shapiro AMJ, Lakey JRT, Ryan EA, Korbutt GS, Toth E, Warnock GL, Kneteman NM, Rajotte RV (2000) Islet transplantation in seven patients with type I diabetes mellitus using a glucocorticoid-free immunosuppressive regimen. New Engl J Med 343:230–238

Shatos MA, Mizumoto K, Mizumoto H, Kurimoto Y, Klassen H, Younng MJ (2001) Multipotent stem cells from the brain and retina of green mice. (E-biomed) J Reg Med 2:13–15

Shors T, Miesegaes G, Beylin A, Zhao M, Rydel T, Gould E (2001) Neurogenesis in the adult is involved in the formation of trace memories. Nature 410:372–375

Sicard RE, Nguyen MP (1994) Interstitial fluids associated with wound repair support proliferation but not differentiation of neonatal rat myoblasts in vitro. Wound Rep Reg 2:306–313

Sicard RE, Nguyen LMP, Witzke JD (1997) Mamalian wound repair environment does not permit skeletal muscle regeneration. Wound Rep Reg 5:39–46

Simpson SB Jr (1983) Fasciculation and guidance of regenerating central axons by the ependyma. In: Kao CC, Bunge RP, Reier PJ (eds) Spinal cord reconstruction. Raven Press, New York, pp 151–162

Sincropi DV, McIlwain DL (1983) Changes in the amounts of cytoskeletal proteins within the perikarya and axons of regenerating frog motoneurons. J Cell Biol 96:240–247

Sirica AE (1995) Ductular hepatocytes. Histol Histopathol 10:433–456

Slack M (1995) Developmental biology of the pancreas. Development 121:1569–1580

Slack JMW (2000) Stem cells in epithelial tissues. Science 287:1431–1433

Smart I (1961) The subependymal layer of the mouse brain and its cell production as shown by radioautography after thymidine-H^3 injection. J Comp Neurol 116:325–348

Smith A (2001) Embryonic stem cells. In: Marshak DR, Gardner R, Gottlieb D (eds) Stem cell biology. Cold Spring Harbor Laboratory Press, Cold Spring Harbor, pp 205–230

Smith CK, Janney MJ, Allen RE (1994) Temporal expression of myogenic regulatory genes during activation, proliferation, and differentiation of rat skeletal muscle satellite cells. J Cell Physiol 159:379–3885

Song K, Wang Y, Sassoon D (1992) Expression of *Hox-7.1* in myoblasts inhibits terminal differentiation of and induces cell transformation. Nature 360:477–481

Song H, Stevens CS, Gage FH (2002) Astroglia induce neurogenesis from adult neural stem cells. Nature 417:39–44

Soonpaa MH, Koh GY, Klug MG, Field LJ (1994) Formation of nascent intercalated discs between grafted fetal cardiomyocytes and host myocardium. Science 264:696–698

Soon-Shiong P, Heintz R, Meredith N, Yao Q, Yao Z, Zheng T, Murphy M, Moloney M, Schmehl M, Harris M, Mendez R, Sandford P (1994) Insulin independence in a type 1 diabetic patient after encapsulated islet transplantation. Lancet 343:950–951

Spangrude G, Heimfeld S, Weissman IL (1988) Purification and characterization of mouse hematopoietic stem cells. Science 241:58–62

Sperry RW (1944) Optic nerve regeneration with return of vision in anurans. J Neurophysiol 7:57–69

Steinman L (2001) Multiple sclerosis: a two-stage disease. Nature Immunol 2:762–765

Stensaas LJ (1983) Regeneration in the spinal cord of the newt *Notophthalmus (Triturus) pyrrhogaster.* In: Kao CC, Bunge RP, Reier PJ (eds) Spinal cord reconstruction. Raven Press, New York, pp 121–149

Stensaas LJ, Feringa ER (1977) Axon regeneration across the site of injury in the optic nerve of the newt *Triturus pyrrhogaster.* Cell Tissue Res 179:501–516

Steward O, Schauwecker PE, Guth L, Zhang Z, Fujiki M, Inman D, Wrathall J, Kempermann G, Gage FH, Saatman KE, Raghupathi R, McIntosh T (1999) Genetic approaches to neurotrauma research: opportunities and potential pitfalls of murine models. Exp Neurol 157:19–42

Stockwell RA (1979) Biology of Cartilage Cells. Cambridge University Press, Cambridge, pp 213–240

Stocum DL (1995) Wound repair, regeneration, and artificial tissues. RG Landes, Austin, pp 1–230

Stocum DL (1996) A conceptual framework for analyzing axial patterning in regenerating urodele limbs. Int J Dev Biol 40:773–784

Stocum DL (1998) Regenerative biology and engineering: strategies for tissue restoration. Wound Rep Reg 6:276–290

Stocum DL (2001) Stem cells in regenerative biology and medicine. Wound Rep Reg 9:429–442

Stocum DL (2004) Amphibian regeneration and stem cells. In: Heber-Katz E (ed) Regeneration: stem cells and beyond. Springer-Verlag, Heidelberg, pp 1–70

Stokes BT, Fox P, Hollinden G (1983) Extracellular calcium activity in the injured spinal cord. Exp Neurol 80:561–572

Sullivan KM, Lorenz HP, Adzick NS (1993) The role of transforming growth factor-beta in human fetal wound healing. Surg Forum 44:625–627

Suzuki K, Suzuki Y, Tanihara M, Ohnishi K, Hashimoto T, Endo K, Nishimura Y (2000) Reconstruction of rat peripheral nerve gap without sutures using freeze-dried alginate gel. J Biomed Mater Res 49:528–533

Svendsen CN, Caldwell MA (2000) Neural stem cells in the developing central nervous system: implications for cell therapy through transplantation. Prog Brain Res 127:13–34

Tanaka EM (2003) Regeneration: if they can do it, why can't we? Cell 113:559–562

Tatsumi R, Anderson E, Nevoret CJ, Halevy O, Allen RA (1998) HGF/SF is present in normal adult skeletal muscle and is capable of activating satellite cells. Dev Biol 194:114–128

Taub H (1996) Transcriptional control of liver regeneration FASEB J 10:413–427

Tavassoli M, Crosby WH (1968) Transplantation of marrow to extramedullary sites. Science 161:548–556

Taylor DA, Atkins BZ, Hungspreugs P, Jones TR, Reedy MC, Hutcheson KA, Glower DD, Kraus WE (1998) Regenerating functional myocardium: improved performance after skeletal myoblast transplantation. Nature Med 4:929–933

Taylor SI (1999) Deconstructing type 2 diabetes. Cell 97:9–12

Taylor G, Lehrer MS, Jensen PJ, Sun T-T, Lavker RM (2000) Involvement of follicular stem cells in forming not only the follicle but also the epidermis. Cell 102:451–461

Temple S (2001) The development of neural stem cells. Nature 414:112–117

Terada N, Hamazaki T, Oka M, Hoki M, Mastalerz DM, Nakano Y, Meyer EM, Morel L, Petersen BE, Scott EW (2002) Bone marrow cells adopt the phenotype of other cells by spontaneous cell fusion. Nature 416:542–545

Thiese ND, Saxena R, Portmann BC, Thung SN, Yee H, Chiriboga L, Kumar A, Crawford JM (1999) The canals of Hering and hepatic stem cells in humans. Hepatology 30:1425–1433

Theise ND, Nimmakayalu M, Gardner R, Illei PB, Morgan G, Teperman L, Henegariu O, Krause D (2000) Liver from bone marrow in humans. Hepatology 32:11–16

Thiese ND, Badve S, Saxena R, Henegariu O, Sell S, Crawford JM, Krause DS (2000) Derivation of hepatocytes from bone marrow cells in mice after radiation-induced myeloablation. Hepatology 31:235–240

Thomson JA, Itskovitz-Eldor J, Shapiro SS, Waknitz MA, Swiergiel JJ, Marshall VS, Jones JM (1998) Embryonic stem cell lines derived from human blastocysts. Science 282:1145–1147

Thorgeirsson S (1996) Hepatic stem cells in liver regeneration. FASEB J 10:1249–1256

Till JE, McCulloch EA (1980) Hemopoietic stem cell differentiation. Biochim Biophys Acta 605:431–459

Tomita S, Li RK,Meisel RD, Mickle DA, Kim EJ, Tomita S, Jia ZQ, Yau TM (1999) Autologous transplantation of bone marrow cells improves damaged heart function. Circulation 100:II247–II256

Torchinsky C et al (1999) Regulation of p27Kip1 during gentamicin mediated hair cell death. J Neurosci 28:913–924

Tobias DA, Dhoot NO, Wheatley MA, Tessler A, Murray M, Fischer I (2001) Grafting of encapsulated BDNF-producing fibroblasts into the injured spinal cord without immune suppression in adult rats. J Neurotrauma 18:287–301

Trachtenberg JT, Chen BE, Knott GW, Feng G, Sanes JR, Welker E, Svoboda K (2002) Long-term in vivo imaging of experience-dependent synaptic plasticity in adult cortex. Nature 420:788–794

Travis J (1994) Glia: the brain's other cells. Science 266:970–972

Trembly JH, Steer CJ (1998) Perspectives on liver regeneration. J Minn Acad Sci 63:37–46

Tropepe V, Coles BLK, Chiasson BJ, Horsford D, Elia J, McInnes RR, van der Kooy D (2000) Retinal stem cells in the adult mammalian eye. Science 287:2032–2036

Turner JT, Singer M (1974) The ultrastructure of regeneration in the severed newt optic nerve. J Exp Zool 190:249–288

Urist MR (1965) Bone: formation by autoinduction. Science 150:893–899

Urist MR, MacLean FC (1952) Osteogenic potency and new bone formation by induction in transplants to the anterior chamber of the eye. J Bone Joint Surg 34A:443–446

Vacanti MP, Roy A, Cortiella J, Bonassar L, Vacanti CA (2001) Identification and initial characterization of spore-like cells in adult mammals. J Cellular Biochem 80:455–460

van Praag H, Kempermann G, Gage FH (1999a) Running increases cell proliferation and neurogenesis in the adult mouse dentate gyrus. Nat Neurosci 2:266–270

van Praag H, Christie BR, Sejnowski TJ, Gage FH (1999b) Running enhances neurogenesis, learning and long-term potentiation in mice. Proc Natl Acad Sci USA 96:13427–13431

van Praag H, Schinder AF, Christie BR, Toni N, Palmer T, Gage FH (2002) Functional neurogenesis in the adult hippocampus. Nature 415:1030–1034

Vassilopoulos G, Wang P-R, Russell DW (2003) Transplanted bone marrow regenerates liver by cell fusion. Nature 422:901–904

Vermeer PD, Einwalter LA, Moninger TO, Rokhlina T, Kern JA, Zabner J, Welsh MJ (2003) Segregation of receptor and ligand regulates activation of epithelial growth factor receptor. Nature 422:322–326

Vortkamp A (2001) Interaction of growth factors regulating chondrocyte differentiation in the developing embryo. Osteoarth Cartilage 9 (Suppl A):S109–S117

Vortkamp A, Lee K, Lanske B, Segre GV, Kronenberg H, Tabin CJ (1996) Regulation of rate of cartilage differentiation by Indian hedgehog and PTH-related protein. Science 273:613–622

Wagers AJ, Sherwood RI, Christensen JL, Weissman IL (2002) Little evidence for developmental plasticity of adult hematopoietic stem cells. Science 297:2256–2259

Wagner W, Reichl J Wehrmann M, Zenner H-P (2001) Neonatal rat cartilage has the capacity for tissue regeneration. Wound Rep Reg 5:531–536

Wakabayashi Y, Komori H, Kawa UT, Mochida K, Takahashi M, Qi M, Otake K, Shinomiya K (2001) Functional recovery and regeneration of descending tracts in rats after spinal cord transection in infancy. Spine 26:1215–1222

Wakitani S, Goto T, Pineda SJ, Young RG, Mansour JM, Caplan AI, Goldberg VM (1994) Mesenchymal cell-based repair of large, full-thickness defects of articular cartilage. J Bone Joint Surg 76:579–592

Wakitani S, Saito T, Caplan A (1995) Myogenic cells derived from rat bone marrow mesenchymal stem cells exposed to 5-azacytidine. Muscle Nerve 18:1417–1426

Wakayama T, Tabar V, Rodriguez I, Perry ACF, Studer L, Mombaerts P (2001) Differentiation of embryonic stem cell lines generated from adult somatic cells by nuclear transfer. Science 292:740–742

Wamil AW, Wamil BD, Hellerqvist CG (1998) CM101-mediated recovery of walking ability in adult mice paralyzed by spinal cord injury. Proc Natl Acad Sci USA 95:13188–13193

Wang S, Shum-Tim D, Galipeau J, Chedrawy E, Eliopoulos N, Chiu RJ-C (2000) Marrow stromal cells for cellular cardiomyoplasty feasibility and potential clinical advantages. J Thoracic Surg 120:999–1006

Wang X, Willenbring H, Akkari Y, Yorimaru Y, Foster M, Al-Dhalimy M, Lagasse E, Finegold M, Olson S, Grompe M (2003) Cell fusion is the principal source of bone-marrow-derived hepatocytes. Nature 422:897–901

Wanner M, Lang DM, Bandtlow CE, Schwab ME, Bastmeyer M, Steuermer CA (1995) Reevaluation of the growth-permissive substrate properties of goldfish optic nerve myelin and myelin proteins. J Neurosci 15:7500–7508

Weidner N, Blesch A, Grill RJ, Tuszynski MH (1999) Nerve growth factor-hypersecreting Schwann cell grafts augment and guide spinal cord axonal growth and remyelinate central nervous system axons in a phenotypically appropriate manner that correlates with expression of L1. J Comp Neurol 413:495–506

Weiss SM, Dunne C, Hewson J, Wohl C, Wheatley M, Peterson AC, Reynolds BR (1996) Multipotent CNS stem cells are present in the adult mammalian spinal cord and ventricular neuraxis. J Neurosci 16:7599–7609

Weissman IL (2000) Stem cells: units of development, units of regeneration, and units in evolution. Cell 100:157–168

Werner S, Smola H, Liao X, Longaker MT, Krieg T, Hofschneider PH, Williams LT (1994) The function of KGF in morphogenesis of epithelium and reepithelialization of wounds. Science 266:819–822

Westerfield M (1987) Substrate interqactions affecting motor growth cone guidance during development and regeneration. J Exp Biol 132:161–176

Wewer UM, Iba K, Durkin ME, Nielsen FC, Loechel F, Gilpin BJ, Kuang W, Engvall E, Albrechtsen R (1998) Tetranectin is a novel marker for myogenesis during embryonic development, muscle regeneration, and muscle cell differentiation in vitro. Dev Biol 200:247–259

Whitby DJ, Ferguson MWJ (1991) The extracellular matrix of lip wounds in fetal, neonatal and adult mice. Development 112:651–668

Whitby DJ, Ferguson MWJ (1991) Immunohistochemical localization of growth factors in fetal wound healing. Dev Biol 147:207–215

Wilmut I, Beaujean N, de Sousa A, Dinnyes A, King T, Paterson LA, Wells DN, Young LE (2002) Somatic cell nuclear transfer. Nature 419:583–586

Woerly S, Pinet E, deRobertis L, VanDeip D, Bousmina M (2001) Spinal cord repair with PHPMA hydrogel containing RGD peptides (NeuroGel). Biomater 22:1095–1111

Wolfe D, Nye HLD, Cameron J (2000) Extent of ossification at the amputation plane is correlated with the decline of blastema formation and regeneration in *Xenopus laevis* hindlimbs. Dev Dyn 218:681–697

Woloshin P, Song K, Degnin A, Killary DJ, Goldhamer DJ, Sassoon D, Thayer MJ (1995) MSX1 inhibits MyoD expression in fibroblast X 10T1/2 cell hybrids. Cell 82:611–620

Wong M, Kireeva ML, Kolesnikova TV, Lau LF (1997) Cyr61, product of a growth factor-inducible immediate-early gene, regulates chondrogenesis in mouse limb bud mesenchymal cells. Dev Biol 192:492–508

Woodbury D, Schwarz EJ, Prockop DJ, Black IB (2000) Adult rat and human bone marrow stromal cells differentiate into neurons. J Neurosci Res 61:364–370

Woodley DT (1996) Reepithelialization. In: Clark RAF (ed) The molecular and cellular biology of wound repair. Plenum Press, New York, pp 339–354

Wurst W, Bally-Cuif L (2001) Neural plate patterning upstream and downstream of the isthmic organizer. Nat Rev Neurosci 2:99–108

Yablonka-Reuvini Z, Quinn LS, Nameroff M (1987) Isolation and clonal analysis of satellite cells from chicken pectoralis muscle. Dev Biol 119:252–259

Yamaguchi Y, Mann DM, Ruoslahti E (1990) Negative regulation of transforming growth factor-β by the proteoglycan decorin. Nature 346:282–284

Yang LJ, Jin Y (1990) Immunohistochemical observations on bone morphogenetic protein in normal and abnormal conditions. Clin Orthopaed 257:249–256

Yang L, Li S, Hatch H, Ahrens K, Cornelius J, Petersen BE, Peck AB (2002) In vitro trans-differentiation of adult hepatic stem cells into pancreatic endocrine hormone-producing cells. Proc Natl Acad Sci USA 99:8078–8083

Yannas IV (2001) Tissue and organ regeneration in adults. Springer, New York

Yaszemski J, Payne R, Hayes WC, Langer S, Aufdemotte TB, Mikos AG (1995) The ingrowth of new bone tissue and initial mechanical properties of a degrading polymeric composite scaffold. Tissue Eng 1:41–52

Yednock TA, Cannon C, Fritz LC, Sanchez-Madrid F, Steinman L, Karin N (1992) Nature 356:63–66

Ying Q-L, Nichols J, Evans EP, Smith AG (2002) Changing potency by spontaneous fusion. Nature 416:545–548

Yoon Ji-Won, Yoon C, Lim H-W. Huang QQ, Kang Y, Pyun KH, Hirasawa K, Sherwin RS, Jun H-S (1999) Control of autoimmune diabetes in NOD mice by GAD expression or suppression in β cells. S131

Young RG, Butler DL, Weber W, Caplan AI, Gordon SL, Fink DJ (1998) Use of mesenchymal stem cells in a collagen matrix for Achilles tendon repair. J Ortho Res 16:406–413

Young HE, Steele A, Bray RA, Detmer K, Blake, Lucas PW, Black AC (1999) Human pluripotent and progenitor cells display cell surface cluster differentiation markers C10, C13, C56, and MHC class-I. Proc Soc Exp Biol Med 221:63–71

Zakon H, Capranica RR (1981) Reformation of organized connections in the auditory system after regeneration of the eighth nerve. Science 213:242–244

Zamora A.J. and Mutin M (1988) Vimentin and glial fibrillary acidic protein in radial glia of the adult urodele spinal cord. Neurosci 27:279–288

Zhang FC, Clarke JDW, Ferretti P (2000) FGF-2 up-regulation and proliferation of neural progenitors in the regenerating amphibian spinal cord in vivo. Develop Bio 225:381–391

Zhou FC, Chiang YH (1998) Long-term nonpassaged EGF-responsive neural precursor cells are stem cells. Wound Rep Reg 6:337–348

Zhou FC, Kelley MR, Chiang YH, Young P (2000) Three- to four-year-old nonpassaged EF-responsive neural progenitor cells: proliferation, apoptosis, and DNA repair. Exp Neurol 164:200–208

Zhou Y, Baumgartner BJ, Hill-Felberg S, McGowen LR, Shine HD (2003) Neurotrophin-3 expressed in situ induces axonal plasticity in the adult injured spinal cord. J Neurosci 23:1424–1439

Ziegler B, Valtieri M, Porada GA, Maria R, Muller R, Masella B, Gabbianelli M, Casella I, Pelosi E, Bock T, Zanjani ED, Peschle C (1999) KDR receptor: a key marker defining hematopoietic stem cells. Science 285:1553–1558

Zou H, Wieser R, Massague J, Niswander L (1997) Distinct roles of type I bone morphogenetic protein receptors in the formation and differentiation of cartilage. Genes Dev 11:2191–2203

Subject Index

A
adult stem cell   2
Axon Regeneration   9, 14

B
biomaterial   56
Blood   36
Bone   33
bone marrow reconstitution assay   38
Bone Regeneration   34

C
Cartilage and Bone Injury   55
Cell Transplant   47
chimeric embryo assay   38
Compensatory hyperplasia   2

D
dedifferentiation   2
Demyelinating Disorder   49
dermal papilla   7
dermal regeneration   66
Developmental Potential   37
Diabetes   53

E
Early Vs Late Stage Frog Limb Bud   72
EGF   24
Epidermis   4

F
FACS   19
Fetal Vs Adult Mammalian Cartilage   71
Fetal Vs Adult Mammalian Skin   69
Fibronectin   11
fumarylacetoacetate hydrolase   22

G
GFP   19
glial scar   12
glutamate toxicity   12
growth cone   11
Growth Factor   34

H
hair follicle   7
hepatocyte growth factor   8
heregulin   6
hippocampus   18–19
HSCs   36

I
IL-6   24
Ischemic injury   21

L
liver   21

M
Mesenchymal Stem Cell   33
MRFs   32
MRL/lpr mouse   68
multipotent adult progenitor cell   43
multipotent muscle stem cell   30
Muscle Regeneration   30
myelin protein   13
Myocardial Infarction   51

N
Neocortex   20
neuroprotective and survival agent   61
neurotrophic factor   11
NF-κB   24

O
Olfactory Bulb   18
olfactory ensheathing cell   63
olfactory ensheathing glial cell   12
Olfactory Nerve   11
optic nerve   13
Osteogenesis Imperfecta   53
oval cell   27
Oval stem cell   39

P
Pancreas   27
Parkinson's Disease   50
partial hepatectomy   22

pluripotency   4
prospective potency   38
prospective significance   37
PwDlx-3   17

R
regeneration-competent cell   3
Regenerative Medicine   47
Retinoic acid   15

S
Satellite cell   30
Schwann cell   10
Skeletal Muscle   29
Skin Injury   54
small intestine submucosa   67
source of cell   55
Spinal Cord   12
Spinal Cord Injury   50

Spinal Nerve   9
STAT3   24
striatum   18
subventricular zone   19
synaptic plasticity   64

T
Tail regeneration   16
TGF-α 24
TGF-β 6
transdifferentiation   38
Tubular nerve guide   60

V
VEGF-A   25

W
Wallerian degeneration   10
Wnt-10   17